EVERYDAY CULTURE IN EUROPE

Everyday Culture in Europe

Approaches and Methodologies

Edited by

MÁIRÉAD NIC CRAITH
University of Ulster

ULLRICH KOCKEL
University of Ulster

REINHARD JOHLER
Eberhard Karls University of Tübingen, Germany

LONDON AND NEW YORK

First published 2008 by Ashgate Publishing

2 Park Square, Milton Park, Abingdon, Oxon OX14 4RN
711 Third Avenue, New York, NY 10017, USA

Routledge is an imprint of the Taylor & Francis Group, an informa business

First issued in paperback 2016

British Library Cataloguing in Publication Data
Everyday culture in Europe : approaches and methodologies.
- (Progress in European ethnology)
1. Ethnology - Europe 2. Europe - Politics and government
I. Nic Craith, Máiréad II. Kockel, Ullrich III. Johler, Reinhard
305.8'0094

Library of Congress Cataloging-in-Publication Data
Everyday culture in Europe : approaches and methodologies/ by Máiréad Nic Craith, Ullrich Kockel and Reinhard Johler.
p. cm.
Includes index.
ISBN 978-0-7546-4690-7
1. Ethnology--Europe. 2. Europe--Politics and government. I. Kockel, Ullrich. II. Nic Craith, Máiréad. III. Title.

D1056.J635 2008
305.80094--dc22

2008010701

ISBN 978-0-7546-4690-7 (hbk)
ISBN 978-1-138-26443-4 (pbk)

Contents

List of Contributors

Bojan Baskar is Professor of Social Anthropology and Mediterranean Studies at the Ethnology and Cultural Anthropology Department of the University of Ljubljana. His current research interests include Mediterranean ethnoecology, nationalism and cultural racism, European regionalism, construction of cultural boundaries, Balkanist discourse, imperial legacies and travel. He is the author of *Ambiguous Mediterranean: Studies in Regional Superposition* (2002, in Slovene).

Anthony D. Buckley is a Senior Honorary Research Fellow in Social Anthropology at Queen's University, Belfast. Formerly, he was a curator at the Ulster Folk and Transport Museum in Cultra, County Down, where he collected, researched and exhibited aspects of community life of Northern Ireland. His research interests have included folk medicine, ethnicity, oral history, religion, and secretive brotherhoods, including friendly societies, the Orange Order and Freemasonry. He is currently exploring the relation between theatre, ritual and sport. His first degree from York University was in history; his MA in religious studies from Leicester University included a dissertation on the *aláàdúrà* churches of Nigeria; his doctorate from Birmingham University, was concerned with traditional medicine in Western Nigeria. Publications include *Yoruba Medicine; A Gentle People; Negotiating Identity; Symbols in Northern Ireland; Border-Crossing, Mumming in Cross-community Contexts; Brotherhoods in Ireland; Unofficial Healing in Ulster; Aristotle and Cricket.*

Vytis Ciubrinskas PhD is Associate Professor of Social Anthropology at the Department of Sociology and Head of the Centre of Social Anthropology at Vytautas Magnus University in Kaunas, Lithuania. He also teachers anthropology at Vilnius University and is an adjunct Associate Professor at Southern Illinois University in the USA. He has been awarded fellowships from the British Council, Fulbright and the Helsinki Collegium for Advanced Studies. He is an editor of *Lithuanian Ethnology, Studies in Social Anthropology and Ethnology*, and is on the editorial board of *Anthropological Journal of European Cultures.* His major research interests are in identity politics, transnationalism and transmigration in Europe and across the Atlantic. His publications include a volume on the ethnography and ethnohistory of the pre-reformed Russian Orthodox ethno-confessional minority in Lithuania. His book on anthropological theory was recently published. He has also published numerous book chapters and journal articles on Lithuanian migration and ethno-nationalism from New-European and North American perspectives.

Ullrich Kockel has worked in higher education in Britain, Germany and Ireland since 1984. He currently holds the Chair of Ethnology and Folk Life at the University of Ulster, Northern Ireland, and is Visiting Professor at the University of the West of England, Bristol. His research ranges across the field of European ethnology, with special focus on cultural encounters (*Borderline Cases: The Ethnic Frontiers of European Integration* 1999), and on culture and economy (*Regional Culture and Economic Development: Explorations in European Ethnology* 2002). In 2003, he was elected to the United Kingdom's Academy for the Social Sciences. He has co-edited a number of recent books including *Negotiating Culture: Moving and Mixing in Contemporary Europe*, 2007 (with Reginald Byron) and *Cultural Heritages as Reflexive Traditions*, 2007 (with Máiréad Nic Craith).

Orvar Löfgren is Professor of European Ethnology at the University of Lund, Sweden. His research is focused on the cultural analysis of everyday life. He has written on national identities and transnational processes, as well as culture, economy and consumption. He is currently in charge of a research project on 'the cultural dynamics of the inconspicuous', focusing on the analysis of mundane routines and non-events, from waiting to daydreaming. Among his recent books are *Off the Edge: Experiments in Cultural Analysis* (edited with Richard Wilk, 2006), *Magic, Culture and the New Economy* (edited with Robert Willim 2005) and *On Holiday: A History of Vacationing* (2000).

Máiréad Nic Craith is Professor of European Culture and Society at the University of Ulster, Northern Ireland. She has previously been attached to the University of Liverpool and University Colleges, Dublin and Cork. She is author and editor of ten books including *Language, Power and Identity Politics* (2007) and *Europe and the Politics of Language* (2006). Her research interests include culture and identity politics, auto-ethnographies and European ethnology. She was joint winner of the 2004 Ruth Michaelis-Jena Ratcliff research prize for folklife. In 2006, she was awarded a Senior Distinguished Research Fellowship at the University of Ulster. She was joint winner of the McCrea Literary Prize in 2008.

Bjarne Rogan holds a MA in French linguistics and a PhD in history. Since 1993, he has been Professor of European Ethnology at the University of Oslo, where he is also Professor of Cultural History. His extensive publications include ten books and some 100 articles in fields such as material culture and consumption studies, maritime anthropology (littoral communities and North Sea fisheries), museology and the historiography of European ethnology. He has worked on both historical and contemporary topics, within the period 1700–2000, and he has a broad experience with both archive studies, fieldwork and participant observation. Among his major assignments are several periods as head of university departments, chairman of research centres and museums, Director of French-Norwegian University Centre (Paris), and Dean of Faculty of Humanities at the University of Oslo.

Elka Tschernokoshewa is the Head of the Department of Empirical Cultural Studies and Anthropology at the Sorbian Institut, Bautzen, Germany. Born in Sofia, Bulgaria, she studied cultural studies and aesthetics in Berlin and received her doctorate from Humboldt University Berlin and post doctorate degree from Sofia.

An Associate Professor, she is a visiting Professor of Cultural Studies and Policy at the Universities of Bristol, Basel, Tübingen, Bremen, Berlin, and Sofia. Her main research interests include cultural diversity, minorities, transculturality, gender and comparative studies on Eastern Europe and Western Europe. She has conducted comparative studies on minorities and migration, and in 2000 she completed the study 'The Pure and the Hybrid. The German Medium Press on Other and Otherness with reference to the Sorbs'. Dr. Tschernokoshewa is an active member of the Commission of Experts on Gender Studies, Sachsen, and a board member of the European Association of Cultural Researchers (ECURES). She is the editor of the book series *Hybride Welten* [Hybrid Worlds], Waxmann Verlag, with themed issues on: *Searching Hybrid Stories. Theories – Empirical Studies – Practice* (2005); *Diversity as Routine. The Accommodation of Difference in Education, the Media and Politics* (2001). Elka is co-editor of the book *Stories of Relationships* (2007), and a team member of the Study for the European Commission 'National Approaches to Intercultural Dialogue in Europe' (2007–2008).

Galia Valtchinova graduated in history at the University of Sofia (1983), and then focused on historical ethnography for her doctorate (1988, Sofia). After a period of specialisation with the Paris school of historical anthropology, she turned her attention to ethnology and social anthropology, conducting fieldwork in the border areas in the Balkans. She is a member of the Bulgarian Academy of Sciences, a Senior Research Fellow at the Institute of Thracian Studies, Sofia (full position) and an associate member of the Centre for Social Anthropology in Toulouse (France). She has worked on issues of ethnicity and identity, memory and history in the Balkans, as well as borders and transborder exchange. Since 1999, she has been working on religious revival, nationalism and politics, with special attention to religious visionaries. Along with about fifty articles in English and French, she is the author of two books in Bulgarian, *Balkan Clairvoyants and Prophetesses in the 20th Century* (The University of Sofia Press, 2006) and *Laudae Znepolensia: Local Religion and Identity in Western Bulgaria* (The 'Prof. Marin Drinov' Academic Publisher, 1999). She is also editor of a forthcoming collection of papers on religion and boundaries (The Isis Press).

Chapter 1

From National to Transnational: A Discipline *en route* to Europe

Máiréad Nic Craith

This chapter surveys the field of European ethnology and considers its definition, evolution and contemporary significance. It also examines changes in the discipline particularly at the beginning of the new millennium, with a view to exploring the way forward for those working in the field. Some disciplines, such as history or literature, are easily characterised and have maintained their conceptual integrity over time. 'Clearly defined disciplines are institutionally privileged.' (Bendix 1997: 5) Other fields have had a more difficult birth and are not described with any great precision, particularly if they tend towards interdisciplinarity, as is the case with European ethnology.

Anne Buttimer (1992: 41) an Irish human geographer, in her analysis of the life-cycles of various disciplines, suggests that in the early stages, a newly-emerging field can be characterised as a Phoenix period 'when new life emerges from the ashes with prospects for a fresh beginning'. This emancipatory movement can be a cry for freedom from oppression or an attempt 'to soar to new heights of understanding, being and becoming'. The subsequent 'Faustian' disciplining of the field is a necessary stage in the process of gaining legitimacy in academic circles. Energies are directed towards 'the building of structures, institutions and legal guarantees for their autonomous existence and identity.' (Buttimer 1992: 43) While the Faustian emblem is primarily used by Buttimer as a symbol of support, it is sometimes the case that academic institutions seek to stifle new ideas – particularly in the case of interdisciplinary projects.

I suspect that the notion of ethnology, and in particular European ethnology, is still at the Faustian stage of defining its core canon and its disciplinary boundaries. Chiva (2003: 65) hints at this when he suggests that anthropology/ethnology 'cannot accept that the *ethnological terrain* can be divided, and that its portions be handed over to the care of different cultural or scientific institutions'. On the other hand, it does not accept the view 'that the terrain or any portion thereof is the sole property of its inhabitants. Indeed, this discipline finds no justification for any kind of territorial circumscription' (italics original).

Defining European Ethnology

The concept of ethnology (without the prefix European) is several hundred years old and has been variously defined by different scholars. In the 1830s the renowned French electrician André-Marie Ampére adopted the term to describe: 'the science which studied the places occupied by nations and the races from which they took their origin, the monuments left behind them by their predecessors, the history of their progress and decadence, the religions that they professed.' (Fenton 2004: 18) There are many regional variations of the field of study. Ek (1981: 12–13) defines ethnology as 'the science of people'. He suggests that it is a 'science of the cultures that shape people in different countries and in different socio-economic settings'. It also extends to the 'organization of these socio-economic societies, of the material means with which men work within the framework of their society and the value systems and ideologies that govern their behaviour'. Ethnology is 'the science of the forces that are generated by particular historical situations and which continue to act as seemingly autonomous forces or as parts of inherited structures'. Finally 'it is the science of people and culture which ought to set out and explain cultural differences, and in doing so, actually questions the value systems we have'.

The term ethnology is sometimes conflated (or confused) with ethnography and Valtchinova (this volume) deliberately uses both terms in her chapter title to highlight the ambiguity between them. For Ampére, there was a clear contrast between the two. Ethnology was a comparative and theoretical field whereas ethnography was descriptive. This contrast is accepted in many central European countries. However the distinction has not entirely been accepted in the Soviet Union and many of its Eastern European neighbours, who have different approaches to these concepts. Ciubrinskas (this volume) explains that early twentieth-century Soviet Russia and its eastern neighbours viewed ethnology as a 'bourgeois pseudo-science', preferring instead ethnography which was regarded as a sub-discipline of history. This approach contrasted sharply with the American view which conflates ethnology with cultural anthropology and the British approach which understands ethnology to mean the study of races and languages, and veers towards archaeology.

Much of the confusion regarding the concept ethnology, can be related to the blurred boundaries between it and other fields of inquiry and this is perhaps not that exceptional. Ek (1981: 7) argues that it has been quite logical for ethnological scholars to work with data from neighbouring fields such as history, philology, anthropology, cultural geography etc. 'Since from the start ethnology has had human culture as the object of its research, it has had to work with material of various kinds. And culture and human behaviour cannot be pigeonholed into a single university subject. Or two.' There is a complex relationship between ethnology and history, although Van Gennep attempted to clarify the boundaries between the two when he suggested that ethnology dealt with 'life and contemporary social facts' whereas history dealt 'with dead facts' (Segalen 1996: 148). Any use that ethnologists made of historical resources was primarily to enhance the understanding of 'contemporary social facts'. Yet historians increasingly draw on the field of ethnology, while ethnologists resort to history and to cultural history – 'once a Cinderella among the disciplines.' (Burke 2004: 1)

Chiva (1990) explains that in France, the two terms 'ethnology' and 'history' come together in the notion of *patrimoine ethnologique*, badly translated as 'heritage'. In earlier decades of the twentieth century, the idea of heritage was confined primarily to material culture (to buildings and monuments) but with the passage of time, its terms of reference have broadened to include non-material dimensions such as beliefs, superstitions and folklore. This enhancement of the concept has been strongly endorsed with the signing of the UNESCO Convention for the Safeguarding of the Intangible Cultural Heritage in Paris in 2003.

Some of the confusion regarding the concept of 'ethnology' may also relate to the lack of consistency in the use of the term. One of the first and main labels for the field in Germany was *Volkskunde*. In 1787, that term was introduced by Joseph Mader 'to designate statistical-topographical accounts of rural populations' and the word has been used in that sense for a further two centuries (Kockel 1999: 78). In the early decades of the twentieth century the notion of folklife has also been used for the field now described as ethnology. Folklife research has been described by Stoklund (1983: 8) as 'a grandchild of the Romantic Movement of the nineteenth century'. Initially students of folklife focused on oral lore such as songs, fairy tales and myths. Subsequently however, the emphasis shifted to include material products – houses, furniture, implements etc. Scholars felt a strong need to secure oral and material treasures which were fast disappearing. Two branches of the discipline evolved from the field – folklore and folklife.

Folklife was quickly established as an academic subject in German and Scandinavian countries. In Sweden, for example, a lectureship in folklife was established at Uppsala University in 1909. A decade later, a chair in Nordic and Comparative Folklife Studies was set up in Stockholm. The emphasis in these early decades was very much on material culture – implements of pre-industrial farming etc. The concern with implements of farming and the *ecumene* blurs the boundaries between the ethnologist and the cultural geographer. Estyn Evans (1972: 517), one of the leading lights in folklife studies in Northern Ireland (Buckley, this volume) once suggested that the ethnologist and the cultural geographer are both 'concerned with certain aspects of human culture'. He argued that any differences between them were 'not so much in content as an approach'.

The first volume of the Journal *Folkliv*, published in 1937, contained an essay on material culture (timber construction) as well as a contribution on the folklore of calculating time in Finland. Its editor, Sigurd Erixon, suggested that this new journal filled a gap in offering a publication devoted specifically to the notion of 'European Ethnology'. Erixon used the terms 'Regional Ethnology and *folklivsforskning* interchangeably.' (Jacobsen 2001: 165–66) He suggested that the field of European ethnology had not yet defined its core canon – existing instead as 'ambulating guest with certain related branches of science'. However, he expressed confidence that the journal would enable the discipline to find its place.

In 1972, the titles of academic chairs at Lund, Stockholm and Uppsala were changed from 'Nordic and Comparative Studies' to 'Ethnology'. Folklife and ethnology are still equated and the concept of European ethnology is widely used in many universities in Central Europe today – although each defines it in its own way. The University of Münster, for example, in a website (no longer available) had equated European Ethnology with 'the study of people and their ways of living together in a society in past and present times'. It suggested that the primary elements of the field were 'the social meanings of material and intellectual culture as well as the traditions and transmissions of these cultural forms'.

The Universities of Göttingen and Humboldt (Berlin) highlight the disciplinary context. As European ethnology involves cultural research, the University of Göttingen suggests that it is 'closely related to History, Linguistics, Literature, History of Art, Social and Cultural Anthropology, Anthropology, Religion and more'. The Humboldt University takes a slightly different approach, suggesting that European ethnology 'spans the disciplines of Folklore Studies, Ethnology and History'. It describes the 'promising young discipline' as an 'exciting enterprise'.

Konrad Köstlin, an eminent European Ethnologist, has described the field in terms of the telling of stories. The discipline, he suggests (1996: 174–5), 'explains to people what is their practice. It tells stories, referring to that "own", and creates identity by means of assumptions of authenticity'. Although European ethnologists are not the only story-tellers in the academic community, Köstlin suggest that they are more successful than most: 'because their stories fit into everyday life and into contemporary life strategies'. (1996: 178) Such narratives can effectively allow a community or nation 'to perform itself'. Here Köstlin is describing a discipline in active, rather than in passive, terms; in the context of what is does, rather than what it is.

A Distinctive Methodology

I like to think of European ethnology in terms of methodology rather than in content terms and the distinctiveness of the disciplinary approach is very effectively outlined by Löfgren in this volume. European ethnology, he suggests, is characterised by four aspects. Firstly, it is the discipline of the trivial. Ethnologists have long obsessed with microscopic details and have painstakingly recorded and documented every detail. 'Ethnologists were always travelling around in the countryside, surveying, taking photos and talking with people, and in co-operation with the dialectologists they elaborated a systematic procedure for the collection of recordings with ordinary people as active collaborators.' (Stoklund 1983: 18) Findings were mapped and cultural atlases were developed in several European countries (c.f. Dow 2002, Gailey 1972, Gailey *et al* 1976, Rooijakkers and Muerkens 2000, Weiss 1962, Wildhaber 1972). In Sweden, for example, the Atlas of Swedish folk culture was published in 1957 and the process of map-making was crucial to the discipline (Rogan, this volume).

A second characteristic of much ethnological research is its reliance on fieldwork. This methodology is common to social anthropology and demarcates those fields from other disciplines such as history or sociology. (Hansen 2003: 149–50) In the past, there were clear differences in the approach to fieldwork adopted by anthropologists versus ethnologists. The differences lay in the field as defined by each academic community as well as the regularity and intensity of the participant observation. While such differences are now becoming increasingly blurred, it is historically true to say that ethnologists largely studied their own communities while anthropologists engaged in the observation of the more exotic.

This is probably a reflection of the political context in which academics worked. Frykman and Löfgren (2003: 2–3) argue, broadly-speaking, that in nineteenth-century Europe, anthropological research of the exotic tended to develop in nations with colonies – or primitives outside of state boundaries. In contrast, researchers with little or no colonial conquests tended to focus on the 'primitives within' – on the peasant cultures that were rapidly disappearing within the boundaries of the nation-state. This approach 'dominated in the Scandinavian countries as well as in most of Central Europe, where folklorists and ethnologists salvaged the past and constructed an idealised picture of a traditional, national, peasant culture'.

In Germany this distinctiveness has been marked by the terms *Volkskunde* versus *Völkerkunde.* While the former deals with the primitive culture of European peoples, the latter engages with folk-cultures of non-Europeans. Ethnologists belong to the former group. They study their own. Köstlin (1996: 178) argues that it is not by accident that ethnologists are involved in their own field – i.e. the field they are researching. 'We reflect ourselves as a nation, as workers, as homosexuals, as women, as men, as mothers or fathers, as animal-lovers, as wine-drinkers or beer-preferers, as people who are afraid of dying, as those who have a childhood and reflect it – and every story needs a teller as an authority for the explanation of our own life'.

Such a practice inevitably creates tensions for the practitioner between the desire to become fully integrated into the community that he or she is researching and the necessity of analysing the field. Hansen (2003: 160) argues for a phenomenological approach to fieldwork. He pleads for a shift from participant observation to a focus on shared experience. Ethnologists should note what people do rather than what they say. 'Being there – taking part in the activities and events we want to study – is a way of changing our focus from words to deeds, and letting people be responsible for their own actions'. However this generates the problem of academic distance. Ultimately ethnologists are not just part of the community being researched. They are engaged in a process of reflexivity that involves distancing oneself from the very community of which they are a part.

Frykman and Gilje (2003: 39) suggest that the ethnologist needs to engage with 'double hermeneutics; in order to understand the life-work being researched. 'He or she must first interpret the meaning that is created by the actors by following the gaze of the beholder'. Then the researcher endeavours to put this meaning in the context of the environment in which it is lived. 'Through interviews and fieldwork, a possibility is presented to see how far the actor's eye extends'.

Increasingly the distinction between ethnology and anthropology has weakened and at the turn of the millennium, many anthropologists have come to work on home cultures. Kockel (this volume) highlights the trend in cultural studies (as practiced in Britain) to focus on 'modern, developed, industrial societies' at home. In France, where the distinction between the disciplines does not really exist, Isaac Chiva (2003: 48–49) notes that 'the emergences of an ethnology where the (French) investigator himself belongs to the society he studies occurred late and is closely related to the rise of ethnology of the exotic.' He suggests that 'this ethnology came under the inevitable and powerful influence of pre-existing sister sciences – to the extent of being taken over at times, by sociologists, historians, geographers and linguistics'.

But while folklife research has a lot in common with anthropology, it also has a lot in common with history and a third feature of the ethnological approach (as defined by Löfgren, this volume) is its historical dimension. Indeed in a lecture in Manchester in 1961, Evans-Pritchard suggested that the 'differences' between historians and ethnologists were technical rather than methodological. While the anthropologist quoted facts from his or her fieldwork, the historian engaged in a similar activity with his or her archival sources (Segalen 1996: 149). Moreover, the ethnologist is not solely concerned with current lived experience. While much ethnological research deals with contemporary issues, there is always the potential to combine present perspectives with cultural history.

Ethnologists have an analytical choice – to apply a historical perspective or not. History (and particularly cultural history) is a tool available to the ethnologist to enhance his or her own analysis of contemporary meaning. Stoklund (1983: 18) describes this as a 'backward-facing approach' when ethnologists take surviving elements, practices or traditions as their point of departure. While Löfgren argues that ethnologists always have the possibility to refer to the historical context, Stoklund puts the case more strongly, suggesting that ethnology is primarily a historical discipline. He defines ethnologists as specialists who engage in the accumulation of the history of the minutiae. 'They specialized in settlement patterns, house-building, furniture, folk costumes, traditional food etc. Some were amateurs, but most belonged to the staff of an archive or museum'. They were obsessed with the key markers of a peasant culture which ultimately served as the catalyst for folk museums in Scandinavia – in Stockholm, Oslo and Copenhagen which ultimately led to the establishment of key posts in the field. Historical tensions continue in the field today which also endeavours to shape its perspectives towards modernity and to focus on topics of contemporary significance 'such as identity, authenticity, ethnicity and – again – continuity.' (Köstlin 1996: 170)

In that context one might locate the field of European ethnology between history and anthropology. While early ethnologists were intent on historical reconstructions, the research focus shifted gradually towards contemporary society. 'Only in living communities was it possible to study those processes which generated the social and cultural entities, and only in such cases was it possible to apply participant observation which, alongside interviewing, became the most important field technique.' (Stoklund 1983: 20) The ethnologist combined historical material goods with contemporary fieldwork and applied his or her knowledge of ritual and lore in order to achieve an understanding of the context in which communities survived. 'To the ethnologist the relationship between history and anthropology is not a question of "either-or" but a matter of "both-and".' (Stoklund 1983: 27)

A fourth characteristic of European ethnology as outlined by Löfgren (this volume) is its flexibility. This trait is both an asset and a liability and for much of the nineteenth and twentieth centuries, the work of the ethnologist was used as an ideological tool for nation-building in a European context. As culture and politics became intertwined, the field of ethnology became increasingly linked to concepts of homeland and nation-state (cf. Löfgren 1989, Ó Giolláin 1999, Wilson 1978).

Ethnology and Nation-States in Europe

The emphasis on the culture of the peasantry 'was entirely in keeping with Romantic Nationalism'. For many ethnologists of the nineteenth and early twentieth centuries, 'a nation was 'naturally grown', and the peasants were closer to its spirit, soul, and soil than other social classes.' (Klein 2006: 59) The concept of home or home-district became ideologically charged with notions of nature, roots and traditions (Frykman and Löfgren 1987: 63). The peasant house became a national symbol (Stoklund 1999). In Scandinavia, for example, Hazelius attempted to open the eyes of all Swedes (including the urban middle classes) to the spiritual and material traditions of the peasantry. Through a process of harvesting the material peasant culture, the feelings of all Swedes towards their father/motherland would be awakened. 'Scholars and folklore collectors saw themselves as a rescue team picking their way through a landscape of cultural ruins, where scraps and survivals of traditional life-styles could still be found.' (Frykman and Löfgren 1987: 59)

Peasant heritage infused public debates on culture and the traditional village became the utopian model for a community way of life. In 1891, Skansen, the first open-air museum, was opened in Stockholm. This was an opportunity for all classes of Swedes to roam among reconstructed cottages and, for a limited period, to enjoy the pleasures of a rural life-style. It focused their attention on the primitive and the soil and generated a longing for utopian, rural pleasures. The Skansen experience offered temporary respite from an ever-changing, increasingly material world. It generated a sense of rootedness – of cultural wholeness. Yet, it was not without complications. It was a sense of unity that had to embrace regional differences. 'Maintaining difference was simultaneously an act of unification.' (Klein 2006: 61)

Moreover, this attempt to turn the clock backwards was recreating, not so much the landscape of the peasantry, as 'the myth of the way it was and the dream of the way it ought to be. There was a longing for a secure society with no class conflicts and no outsiders.' (Frykman and Löfgren 1987: 63) There was a desire for a secure, homogenous society that was co-terminous with the boundaries of the nation-state. Hybrid cultures were awkward to deal with and purity (or mono-culturalism) was preferable (Tschernokoshewa, this volume). The end-product of this process was the 'imagined community' (Anderson 1983), a perfect nation with a homogenous, folk-population who shared one distinctive culture, language, history and future (Löfgren 1996: 162).

Ethnology was a process which helped define insiders versus outsiders – us versus them. Folklore and Folklife studies defined the national heritage and highlighted the contrast between one's own heritage and that of one's neighbours. A key feature of that heritage was landscape. 'Real' Norwegian folk culture was found in the remote mountain valleys of Telemark. The Gaeltacht fishing villages on the Western seaboard of Ireland came to symbolise Gaelic Ireland. Finnish folk culture survived in the rural forests of Karelia. A sense of national identity was rooted in the physical soil and in the relationship of the peasantry with the language. This was a common feature of nation-building in many locations. 'Dalarna is comparable to Karelia in Finland, Hardanger in Norway, Appalachia in the United States, Dogon land in Mali, and other "old-fashioned" or "relic areas" far away from a capital.' (Klein 2006: 75)

The relationship between belonging and homeland was no less important in empires that in nation-states. Baskar (this volume) suggests that the experience of belonging at regional and national levels in a nation-state is hardly different from the sense of belonging that occurs at national and international levels. The empire offers an additional layer of belonging – although the relationship between core and periphery can be complicated by a number of factors, including national groups. In an imperial context, the core nation may feel under pressure to legitimise the cultural diversity of the whole and to dominate other national groups with separatist tendencies.

Orvar Löfgren (1993) suggests that nation-states have a 'tool-kit' of symbols. The kit contains stories, dances and costumes as well as flags, monuments and national anthems. Ethnologists contributed to the identification of the former and while a liberal *Volkskunde* once referred to 'a friendly and mostly harmless identity game in "developed" societies', its focus on authenticity was a useful tool in offering arguments against strangers. (Köstlin 1996: 170)

In the mid-twentieth century, ethnology was invoked by Nazism, Fascism and Marxism to endorse ideologies that were intolerant and destructive of identities perceived to sully the purity of the nation-state. Folklife studies were suspected of having 'become dominated by National Socialist ideas during the twelve years of Hitler, or of having been a nationalistic pseudo-science even before Hitler.' (Lesser 1970: 285–86) The desire for racial purity was linked to a search for the authentic. 'Nazi ideology presented racial purity as a means to heal the wounds of the suffering German state'. From Hitler's perspective, the ethnic heterogeneity of Germany was a primary reason for the country's political and economic weakness, 'and he promised to restore a German realm based on a cleansed, and hence strong, German people'. This was the principal argument, he used against the Jews, 'but the urge for Germans to prove the authenticity of their own beliefs may have an even greater motivational force than their wanting of "ethnic" purity.' (Bendix 1997: 163)

Once combined with a racist discourse, the 'association between the *Volkskunde* concept of the nation and the ideology of Nazism became almost inescapable'. Moreover, 'the equation of "Arian" with "German", which *Volkskunde* and ethnology more generally, had been made in a somewhat naïve way, thereby offering the National Socialists a "scientific" basis for their ideology.' (Kockel 1999: 87) Following the division of Germany after World War II, the discipline was strongly influenced and supported by Steinitz – a Finno-Ugric scholar. His prominence in communist circles established his influence in folklife studies not only in Germany, but behind the Iron Curtain – in Albania, Bulgaria, Czechoslovakia, Hungary, Poland, Romania and the Soviet Union (Rogan, this volume).

Ethnology became an important conceptual battlefield and the study of the working-class was supported by a regime whose goals it served. Ethnological research was a political tool and from 1949 onwards, 'almost no scholarly study could be published without sincere or constrained statements of adherence to the Marxist creed, without citations from its representatives, and without reaching a conclusion in conformity with Marxist-Leninist doctrines.' (Dégh 1970: 299) For a number of decades, folklore and folklife thrived. Fieldworkers were funded. New institutions were established and modern equipment was purchased in support of an academic field that legitimated the prevailing political ideology. Research of folk-culture and folk-expressions was perfectly in keeping with the ideologies of Marxism and Leninism. While ethnologists in the communist GDR shifted their focus from peasant traditions to the working class, those in non-communist settings were orientating their research towards modern industrial society. The discipline maintained its research focus on neglected groups and minutiae rather than on formal institutions and structures (Löfgren this volume).

Although folklife and ethnology are no longer closely associated with discredited political ideologies, the discipline is still haunted by its political past. 'More than fifty years after the Third Reich collapsed the history of those twelve years still remains an open wound for the discipline'. Kockel (1999: 89) suggests that 'the concept of *Volk*, which lies at its very heart, has become highly suspect, and the whole discipline with it'. One of the issues for academics in the field is to discern how to recover its concepts and terminology while rejecting and discarding the abuse of these that occurred in the mid-twentieth century. It is also challenged by the erosion of peasant culture and the urbanization of contemporary society. With the demise of peasant conditions, European ethnology faces the challenge of re-defining its field of research and asking whether its methodologies are suited to the investigation of non-rural phenomena and the changes that are taking place in contemporary Europe.

The Comparative Dimension

One of the key assets of European ethnology has been its emphasis on a comparative approach to research – a feature that has been consistently highlighted by the giants of ethnology over time. In 1858, Wilhelm Heinrich Riehl, a professor of statistics, and later cultural history at the University of Munich delivered a lecture entitled '*Die Volkskunde als Wissenschaft*' in which he called for a comparative approach to the study of folkculture based primarily on the methodology of participant observation. Sources from libraries, archives etc. were relegated to second place (Kockel 1999: 81).

At the beginning of the twentieth century, Nils Lithberg, an ethnologist in Stockholm, proposed a methodology which would be 'an internationally comprehensive exploration which sought out the geographical and historically broad connections between cultures long since transformed'. The tools of his trade were 'insignificant cultural products such as peppermills and mortars.' (Ek 1981: 7–8) This divergent approach was again highlighted in 1937, by Sigurd Erixon, who proposed that 'in order to be comparative, the real object of ethnology/folklivsforskning was to be how history, society, and geography have influenced "life;" or, "man himself"' (Jacobsen 2001: 166). Erixon was arguing for a holistic approach to enhance our understanding of societies, cultures and traditions. Linda Dégh places ethnology, 'the comparative study of European folk cultures', in a Hungarian context (1970: 300). From her perspective, the comparative focus was already evident in Hungary where 'cross-cultural analyses explore the place of the Hungarian folk culture in the European context.' (Dégh 1970: 306)

Estyn Evans maintained that it was the comparative dimension to research that saved the cultural geographer and the folklife researcher alike from the dangers of parochialism. Both fields focused 'on a small area of the earth's surface, of personal observation, collection, and fieldwork' which could potentially engender parochialism or a 'narrow nationalism'. In the case of the geographer, the desire 'to know why one region differs from another, is a useful corrective'. In a similar manner, the folklife researcher 'is forced to take into account other regions and other times.' For this reason 'there need be no contradiction between intensive research on a local scale and the universal outlook: international reputations can be built on studies of real situations in local environments.' (Evans 1972: 518–19)

When discussing French ethnology, Bromberger (2003: 2) highlights and analyses three different scales of research; using the analogies of the microscope versus the telescope. Although ethnologists usually locate the field in a very small territory, they can avail of either tool. 'Some use a microscope to study the infinitesimal' while 'others use a telescope from their observation post'. He characterises these methodologies as local versus localised.

In the 1950s, ethnological research in France was primarily local. There were many 'precise studies on the functioning of micro-societies (village communities, groups of families, craft enterprises, etc) and even of individuals.' (Bromberger 2003: 1) In confining him- or herself to this methodological approach, the researcher deprived him- or herself of key aspects of various processes and institutions. The community-based approach was particularly poor when dealing with political processes and economic institutions whose effects reached far beyond the local field. There was also the issue of local power struggles. If viewed solely through a local lens, the researcher could potentially miss valuable opportunities to examine the local in a national context. Bromberger preferred a localised approach 'which is equally grounded in a detailed examination of a limited number of facts', but relies also on 'a comparative approach rather than a study of contextual structure and function in order to try to establish their meanings.' (Bromberger 2003: 22)

Ultimately he advocated a combination of local, localised and large-scale research. 'Can one understand the significance of a piece of clothing?' he asks 'by simply referring to its regional context, or by merely analysing its varied uses in its local environment, or by looking into the meanings attached to it by a group, or even by taking into account its symbolic affinities with other clothing in neighbouring or even distant communities?' A methodology 'that privileges only one perspective and one scale of analysis is bound to fail'. Instead he argued that ethnologists should use new methods 'to rethink these old issues' (Bromberger 2003: 26). Martine Segalen (1996: 153) suggests that French ethnologists have adopted the comparative perspective, and advocates a holistic interdisciplinary approach. 'We must use our imagination when dealing with our topics, not limiting ourselves to single sources of data, and putting this in interdisciplinary approach within our research'.

There has been a renewal of emphasis on the comparative approach in very recent decades which has involved looking out of the national context towards a more European framework, but this raised many issues. What kind of Europe were ethnologists looking towards, and where were its boundaries? Was it a geographical Europe that stretched from the Atlantic Ocean in the West, to the Ural Mountains in the East, from Iceland in the North, to the Mediterranean in the South? Was it a

Euro-centric Europe which focused only on 'Europeans' – i.e. White, Caucasian and Christian to the exclusion of 'non-Europeans'?

In an excellent exploration of the European dimension to ethnology, Klaus Roth (1996) asked a series of questions which are still pertinent for the discipline today. 'What does European ethnology really mean?', he asks. It is generally held that the concept does not simply refer to ethnology in Europe; not does it have as its goal 'to set Europe as a unified "own" against the "alien" non-European world' (Roth 1996: 4), but there is no common agreed definition of the field.

Is it 'simply a synonym for the old names, a neat new label for the well-known ("folkloric", "ethnographic") subject matter?' or has it 'developed into a mere collective name for the "European folklores", thus preserving their national focus'? Drawing on Erixon (1967: 5), Roth asks whether the goal of the discipline is 'to unify disparate, national traditions and methods to create "a uniform European folklife research in systematic form" in order to satisfy "the need of a systematic cooperation within ethnology"?' Or is it a sub-discipline which focuses on Europe as a region in the same way that regional ethnologies examine different continents. Finally, he queries, whether the ethnology of Europe is 'a discipline studying the *European* cultures in their variety and their unity'. However such an approach would engender the risk of Eurocentrism (italics original). Roth's concluding questions address the unique emphasis of the field on a comparative approach to research. 'Isn't European ethnology by its very nature a *comparative* science, a "comparative Ethnology of Europe" and above all a science of cultural relations and influences of Europe, of interdependencies and iterations between groups and peoples?' (italics original in Roth 1996: 3)

Adopting a comparative approach has not been easy. In the 1960s and 1970s, very few Swedish ethnologists actually conducted research outside of Sweden. When ethnologists at Lund began to seriously consider the prefix 'European' they were faced with a number of new issues. As their previous research had adopted a local rather than a localised approach, they had very little research experience of comparative contexts. Moreover, new linguistic skills were required to conduct research in other European fields (Hansen 2003: 152).

Frykman (2003: 169) describes the whole process as a slow turn from ethnology in the different countries of Europe to the ethnology of Europe in different countries. Up to that point, ethnologists had investigated many core issues in local and national contexts. In the 1970s, many had studied society and social structure. Systems of culture dominated ethnological research in the 1980s. The issue of cultural identities came to the fore a decade later and the reflexive turn thereafter.

Not only have the geographical scale and themes changed, there has also been a move from what Frykman calls 'the territorial margins' in Europe to central, political issues. In former times, 'European ethnology harboured a proficiency in the subject of sheep in Balkan hills, transhumance in the Alpine regions, the hunting of birds in the North Atlantic coasts, production of cheese where no-one would have believed it was possible, with the aid of age-old implements still surprisingly in use.' (Frykman 2003: 180–81) Such debates were far removed from the everyday lives of contemporary EU citizens. However, from his perspective, the focus was changing and ethnologists were looking forward as well as backward and turning their disciplinary skills towards central political issues.

The call for a more 'European' ethnology is not new. Some thirty-five years ago, Fenton (2004 [1973]: 27) suggested that in an increasingly globalised world 'Europe is itself a region'. Moreover, 'it is to regions on this kind of scale that the European concept of the scope of regional ethnology should be related'. This was not to suggest that all researchers needed to focus on the macro-level, but it does propose a localised approach in a trans-national context. 'Each nation or region must learn to play its part within the complex that goes to make up European ethnology.'

The call for a more European ethnology continues to be heard. Many university websites reflect a strong European focus. In Austria, the University of Vienna suggests that European ethnology 'provides a modernised view of folklore and folklife studies, bringing it together with new European perspectives and taking part alongside many historical-philological disciplines in the dialogue about the most recent tendencies in international cultural sciences'. Apart from the classical areas of research, the department focuses on 'cultural aspects of present-day dynamics in European societies.'

The European focus is strongly evident on the English language version of the website of the Humboldt-University of Berlin. For students here 'the study of European Ethnology does not intend to impart encyclopaedic knowledge about "Europe", but to accomplish a deeper understanding of cultural processes and their dynamics'. Their research projects focus on contemporary issues in Europe today, and 'topics range from the culture of the Internet to that of the Suomi, from company cultures to the "Techno Scene", from the Russian community in Berlin to the post-war cultural migration of German women to Paris.' The Institute for Cultural Anthropology and European Ethnology at the Johann Wolfgang Goethe Universität in Frankfurt-am-Main suggested in a website (no longer accessible) that 'European Ethnology is projected as a new direction taken in the anthropological investigation of European cultures past and present'. It notes the challenges of globalization for the field.

A similar emphasis on Europe prevails in the study of European ethnology at Ludwig-Maximilians-Universität, in Munich. There, the website of the Institut für Volkskunde/Europäische Ethnologie, defines the field as 'an empirical branch of the humanities concerned with examining and analysing the historical and contemporary cultural phenomena of European societies'. Although the discipline has experienced many changes over time, 'what has remained is an interest in the everyday culture of a wide range of social classes and groups and their ways of life as well as in the processes of change continually taking place in European societies'. The spectrum of research has been 'expanded to include, among other things, a comparative study of European societies – with an increasing focus on global integration'.

Ethnology at Turku University in Finland is 'cultural research which focuses particularly on Europe'. Pekka Leimu (2006) suggests that ethnologists at the university promote the European dimension to research in their teaching. 'Finland is an integral part of Europe, the Finns are Europeans and Finnish folk culture is a regional version of a wider European culture.' Ethnologists at Turku University place special emphasis on the notion of 'Middle Europe' perceiving it as 'a transition zone between the cultures of Eastern and Western Europe, which begins in Finland and continues through the Eastern Baltics and eastern-central Europe to the Balkans'. Leimu sees the role of ethnology as establishing the connections as well as the boundaries, between Finns and other cultural nations of Europe. Cole (2003: 31) endorses the relational notion of ethnology and suggests that the reduction of national antagonisms within the EU has generated 'a rapprochement of national ethnologies and serious efforts to create a European ethnology'. However, he confines the success story to the EU suggesting that 'national ethnologies in countries outside the Community face a more difficult task'.

A review of recent editions of *Ethnologia Europaea* does suggest a much stronger focus on the European context from scholars in the field. A special edition in 1999 was devoted to the question of 'Europe as a Cultural Construction or a Reality' and contained articles on the character of Europe by academics such as Harbsmeier, Johler, Kockel, Roth and Schippers. In 2002, the journal published a special edition on 'Getting Europe into place'. Again this volume included several essays on a trans-national theme such as Johler's 'Local Europe', Hansen's 'Festivals, Spatiality and the New Europe' and a contribution by Reme on Bergen as a European city of culture. More recent special editions featured 'museums and modernity' (Niedermüller and Stoklund 2003) and 'multicultures and cities' (Gösta and Butler 2006).

The book series to which this volume belongs, 'Progress in European Ethnology' also features the trans-national context. The monographs and edited volumes are designed to provide a critical overview of different national/regional traditions (Dow and Bockhorn 2004, Kockel 2002, Margry and Roodenburg 2007 and Rihtman-Auguštin 2004). These books should generate a body of academic literature core to the discipline and ultimately provide a library of European ethnology for scholars in the field throughout Europe.

Conclusion

This essay and the co-editorship of this specific volume has been a personal journey for me. My first research project focused on the peasant way of life on the Blasket Islands of the south-west coast of Ireland at the turn of the twentieth century. My primary resource was the ethnic autobiography of Tomás Ó Criomhthain – affectionately known as the 'Islandman' – and the acquisition of literary skills by a people immersed in oral traditions (Ó Criomhthain 1929), This research was published in Irish Gaelic (Nic Craith 1988). Since then I have developed a research interest in a number of parallel research foci – including language and living patterns among peasant communities in the south of Ireland at the beginning of the twentieth century (Nic Craith 1993), cultural identity and traditions in contemporary Northern Ireland (Nic Craith 2002, 2003) and struggles for cultural legitimacy among minority groups in different regions in Europe (Nic Craith 2006). My research focus has shifted from rural to urban, from peasant to working-class, from Ireland to Europe. All of this is perfectly in keeping with the pattern of research in European ethnology – yet, until recently, I would not have dared call myself an ethnologist – a risk that does not necessarily guarantee that others will recognise me as such.

Perhaps this is one of the issues that needs to be addressed by those spearheading the discipline. Who are the members of this academic community? How do we re-develop and re-define concepts and terminology that have been sullied by political dictators? What is the core canon of contemporary ethnological thought in Europe? What are the key publications in which we engage with one another in academic dialogue? How do we apply our methodology to the primary issues of academic concern at a European level at the beginning of the twenty-first century? How do we convince others of the significance of our discipline? The chapters that follow are a 'snapshot' of historical and contemporary ethnological research in Europe. They establish current as well as potential explorations in the field. The challenge to the reader is to take that research forward in a new and meaningful manner.

Acknowledgement

A special thanks to Dr Loredana Salis for help in preparing this text for publication.

Bibliography

Anderson, B. (1983) *Imagined Communities: Reflections on the Origin and Spread of Nationalism.* London, New York: Verso.

Bendix, R. (1997) *In Search of Authenticity: The Formation of Folklore Studies.* Wisconsin: University of Wisconsin Press.

Bromberger, C. (2003) 'From the Large to the Small: Variations in the Scales and Objects of Analysis in the Recent History of the Ethnology of France' in L. Varadarajan and D. Chevallier (eds), N. Padgaonjar transl., *Tradition and Transmission: Current Trends in French Ethnology: The Relevance for India.* New Delhi: Ayran Books International, pp. 1–42.

Burke, P. (2004) *What is Cultural History?* Cambridge: Polity Press.

Buttimer, A. (1993), *Geography and the Human Spirit.* Baltimore and London: Johns Hopkins University Press.

Chiva, I. (2003) 'Ethnological Heritage: The French Example', in L. Varadarajan and D. Chevallier (eds), N. Padgaonjar transl., *Tradition and Transmission: Current Trends in French Ethnology: The Relevance for India.* New Delhi: Ayran Books International, pp. 43–69.

Cole, J.W. (2003) 'European Ethnology: Eight Theses', in G. Pizza and F. Zerilli (eds), *La Ricerca Antropologica in Romania, a cura di Cristina Papa.* Napoli: Edizioni Scientifiche Italiane, pp. 23–36.

Dégh, L. (1970) 'Ethnology in Hungary', in *East European Quarterly*, 4(3), 293–307.

Dow, J.W. (2002) 'Anthropology: The Mapping of Cultural Traits from Field Data', in *Social Science Computer Review*, 12(4), 479–92.

Dow, J.R. and Bockhorn, O. (2004) *The Study of European Ethnology in Austria.* Aldershot, Burlington: Ashgate.

Ek, S. (1981) 'On Ethnology: Inaugural Lecture', *Ethnologia Scandinavica*, 7–12.

Erixon, S. (1967) 'European Ethnology in Our Time', *Ethnologia Europaea*, 1, 3–11.

Evans, E. (1972) 'The Cultural Geographer and Folklife Research', in R. Dorson (ed.), *Folklore and Folklife*. Chicago and London: University of Chicago Press, pp. 517–33.

Fenton, A. [2004] (1973) 'The Scope of European Ethnology', in A. Dundes (ed.) *Folklore: Critical Concepts in Literary and Cultural Studies: From Definition to Discipline*, 1. Chicago and London: University of Chicago Press, pp. 17–28.

Frykman, J. and Gilje, N. (2003) 'Being There: An Introduction', in J. Frykman and N. Gilje (eds), *Being There: New Perspectives on Phenomenology and the Analysis of Culture*. Lund: Nordic Academic Press, pp. 7–52.

Frykman, J. and Löfgren, O. (2003) *Culture Builders: A Historical Anthropology of Middle-Class Life*. New Brunswick: Rutgers University Press.

Gailey, A. (1972) 'Towards an Irish Ethnological Atlas?', in *Ulster Folklife* 18, 121–5.

Gailey, A., Ó Dannachair, C. and Adams, G.B. (1976) 'Ethnological Mapping in Ireland', *Ethnologia Europaea*, IX(1), 14–34.

Gösta Arvastson, Butler, T. (eds) (2004) *Multicultures and Cities*, special edition of *Ethnologia Europaea*, 34(2).

Hansen, K. (2002) 'Festivals, Spatiality and the New Europe', *Ethnologia Europaea*, 32(2), 19–36.

—— (2003) 'The Sensory Experience of Doing Fieldwork in an "Other" Place' in J. Frykman and N. Gilje (eds), *Being There: New Perspectives on Phenomenology and the Analysis of Culture*. Lund: Nordic Academic Press, pp. 149–67.

Harbsmeier, M. (1999) 'Character, Identity and the Construction of Europe', in *Ethnologia Europaea*, 29(2), 5–12.

Jacobsen, J. (2001) 'Creating Disciplinary Identities: The Professionalization of Swedish Folklife Studies' in A. Dundes (ed.), *Folklore: Critical Concepts in Literary and Cultural Studies*, 1. London and New York: Routledge, pp. 158–74.

Johler, R. (1999) 'Telling a National Story with Europe: Europe and the European Ethnology', in *Ethnologia Europaea*, 29(2), 67–74.

—— (2002) 'Local Europe: The Production of Cultural Heritage and the Europeanisation of Places', in *Ethnologia Europaea*, 32(2), 7–18.

Klein, B. (2006) 'Cultural Heritage, the Swedish Folklife Sphere, and the Others', *Cultural Analysis*, 5, 57–80.

Kockel, U. (1999a) *Borderline Cases: The Ethnic Frontiers of European Integration*. Liverpool: Liverpool University Press.

—— (1999b) 'Nationality, Identity and Citizenship: Reflecting on Drumcree Parish Church', in *Ethnologia Europaea*, 29(2), 97–108.

—— (2002) *Regional Culture and Economic Development: Explorations in European Ethnology*. Aldershot, Burlington: Ashgate.

Köstlin, K. (1996) 'Perspectives of European Ethnology', in *Ethnologia Europaea*, 26 (2), 169–80.

Lesser, P. (1970) 'Ethnology in Germany', in *East European Quarterly*, 4(3), pp. 275–92.

Leimu, P. (2006) 'Finnish Ethnology in the 2000s', Paper presented at the NEFA meeting in Turku, Fennicum, 28 October, http://www.hum.utu.fi/oppiaineet/kansatiede/en/research/ethnology2000.html

Löfgren, O. (1989), 'The Nationalization of Culture', in *Ethnologia Europaea*, 19, 5–23.

—— (1996) 'Linking the Local, the National and the Global: Past and Present Trends in European Ethnology', in *Ethnologia Scandinavica*, 26(2), 157–68.

Margry, P.J. and Roodenburg, H. (eds) (2007) *Reframing Dutch Culture: Between Otherness and Authenticity*. Aldershot, Burlington: Ashgate.

Nic Craith, M. (1988) *An tOileánach Léannta*. Dublin: Clóchomhar Teo.

—— (1993) *Malartú Teanga: an Ghaeilge i gCorcaigh sa Naoú hAois Déag*. Bremen: European Society for Irish Studies.

—— (2002) *Plural Identities: Singular Narratives – the Case of Northern Ireland*. New York, Oxford: Berghahn.

—— (2003) *Culture and Identity Politics in Northern Ireland*. Basingstoke: Palgrave Macmillan.

—— (2006) *Europe and the Politics of Language: Citizens, Migrants, Outsiders*. Basingstoke: Palgrave Macmillan.

Niedermüller, P. and Stoklund, B. (eds) (2003) *Museums and Modernity*, special issue of *Ethnologia Europaea*, 33(1).

Ó Criomhthain, T. (1929) *An tOileánach*, An Seabhac eag. Baile Átha Cliath.

Ó Giolláin, D. (1999) *Irish Folklore: Tradition, Modernity, Identity*. Cork: University College Cork.

Reme, E. (2002) 'Exhibition and Experience of Cultural Identity: The Case of Bergen – European City of Culture', in *Ethnologia Europaea*, 32(2), 37–46.

Rihtman-Auguštin, D. (2004) *Ethnology, Myth and Politics: Anthropologizing Croatian Ethnology*, edited by Jasna Čapo Žmegač. Aldershot, Burlington: Ashgate.

Rooijakkers, G. and Muerkens, P. (2000) 'Struggling with the European Atlas: Voskuil's Portrait of European Ethnology', *Ethnologia Europaea*, 30, 75–95.

Roth, K. (1996) 'European Ethnology and Intercultural Communication', *Ethnologia Europaea*, 26(1), 3–16.

—— (1999) 'Toward "Politics of Interethnic Coexistence": Can Europe learn from Multiethnic Empires', *Ethnologia Europaea*, 29(2), 37–51.

Schippers, T. (1999) 'The Border as a Cultural Idea in Europe', *Ethnologia Europaea*, 29(2), 25–30.

Segalen, M. (1996) 'Reconstructing, Understanding and Inventing Past Cultures: The Complex Dialogue between French Ethnology and History', *Ethnologia Europaea*, 26(2), 147–56.

Stoklund, B. (1983) *Folklife Research: Between History and Anthropology*. Cardiff: National Museum of Wales.

—— (1999) 'How the Peasant House became a National Symbol', *Ethnologia Europaea*, 29(1), 5–18.

UNESCO Convention for the Safeguarding of the Intangible Cultural Heritage (Paris, 17 October 2003) http://unesdoc.unesco.org/images/0013/001325/132540e.pdf.

Weiss, R. (1962) 'Cultural Boundaries and the Ethnographic Map' in P. Wagner, and M. Mikesell (eds), *Readings in Cultural Geography*. Chicago: University of Chicago Press, pp. 62–74.
Wildhaber, R. (1972) 'Folk Atlas Mapping' in R.M. Dorson (ed.) *Folklore and Folklife: An Introduction*. Chicago: University of Chicago Press, pp. 479–96.
Wilson, W. (1978) 'The Kalevala and Finnish Politics', in F.J. Oinas (ed.), *Folklore, Nationalism and Politics*. Ohio: Slavica Publishers Ltd, pp. 51–75.

University Websites

Johann Wolfgang Goethe Universität, Frankfurt am Main, http://www.uni-frankfurt.de/fb/fb09/kulturanthro/index.html (website no longer accessible).
Humboldt University Berlin, http://www2.hu-berlin.de/ethno/english/departm/general.htm.
Ludwig-Maximilians-Universität, Műnchen, http://www.volkskunde.uni-muenchen.de/ueber_uns/portrait/english/index.html.
University of Vienna,
http://euroethnologie.univie.ac.at/fileadmin/user_upload/inst_ethnologie_europ/ECTSenglisch.pdf.
University of Göttingen, http://www.kaee.uni-goettingen.de/english/english.htm.
University of Munster, http://www.uni-muenster.de/Volkskunde/english.htm (website no longer accessible).
University of Turku, http://www.hum.utu.fi/oppiaineet/kansatiede/en/research.

Chapter 2

From CIAP to SIEF: Visions for a Discipline or Power Struggle?

Bjarne Rogan

The third remoulding after World War II of CIAP – *la Commission Internationale des Arts et Traditions Populaires* – started at a small conference in Oslo in August 1961. The process ended three years later in Athens, with the dissolution of CIAP and the founding of SIEF – *la Société Internationale d'Ethnologie et de Folklore*. In 1961, a restructuring of this international organization was long overdue. There was some activity in some of its working groups or 'commissions'. But CIAP itself, even as an umbrella organization, was passive and weak, hardly capable of securing funding or of supporting its ongoing projects, not to speak of coming up with new projects. There had been no congress since 1955, the Board meetings were rare, and there was almost no communication with the membership, as both the scientific journal and the newsletter had ceased to appear in the mid 1950s.[1] Except for some specified projects funded by UNESCO, CIAP depended upon the members for its subsistence. But with members unwilling to open the purse and hardly any administration to collect the fees due, it was a vicious circle. The international cooperation that still lingered on, took place in the periphery, and CIAP was no catalyst for further cooperation.

The reorganization was a painful process, full of antagonism and conflicts. As with many conflicts in Academia, it is sometimes difficult to see what was at stake for the disciplines and for the scholarly contents of international cooperation, as so much of the debate was on organizational questions. What emerges clearly from the sources, however, is that the battle of CIAP was not only a striving for the development of European ethnology. It was also, and at periods perhaps even more, a striving for power and control – to the extent that these factors are possible to separate. Many of the discipline's inherent challenges came to the fore, and several of its most important scholars participated – some visibly, some behind the scenes. This chapter focuses on the reorganization of CIAP in the early 1960s. A retrospective glance at the preceding period will serve as a backdrop, and there will be a short

1 The same problem applies to the archive situation. There is no central archive for CIAP. The archives more or less 'belonged' to those in charge – the elected officers. Documents concerning CIAP (and SIEF) are spread around in different archives; at university institutes, in museums and libraries. Official documents, like minutes, Reports and newsletters, are relatively few in number and give only a limited insight into the history of CIAP and SIEF. But for the internal life of the organization, or for the lack of such, and for the strategies of its central members, correspondence and private memos are an important source.

presentation of the subsequent events – as the belligerent parties by no means laid down arms after Athens. Among other things, the journal *Ethnologia Europaea* was founded in overt opposition to the reconstructed SIEF.

The Backdrop: Post-war CIAP – from Optimism to Lethargy

CIAP had been founded in 1928 in Prague during a folk art congress organised by the League of Nations. During the 1930s, CIAP was under the strict control of the League's commission for cultural cooperation, for its administrative as well as its scientific activities.[2] But the declining prestige and influence of the League itself in the interwar years were detrimental to its sub organizations, CIAP included. When first Germany withdrew from the League (1933) and later Italy (1937), CIAP was paralyzed.

Postwar life in CIAP started on an optimistic pitch. After a preliminary meeting in Geneva in 1945 and a General Assembly in Paris in 1947, CIAP was given a new start – with a new bylaw and a new structure. At the 1947 conference in Paris, the around sixty delegates boiled over with enthusiasm. There was a unanimous will to be 'a strictly scientific organization, without the intervention of governmental authorities' and to escape all the traps that the old CIAP had fallen into. At the same time there was an unrealistic optimism about activities to be started, the creation and recreation of ethnological institutions after the war, the use of the discipline to reconstruct the rural zones of Europe, and so on.[3]

In 1947, CIAP had started out on its own, with an independent status in relation to UNESCO. But with no funding there could be no activities, so a change of policy was necessary. In 1949, CIAP joined a group of international scientific organizations to found the UNESCO organism CIPSH – *le Conseil International de Philosophie et des Sciences Humaines*. As a member of CIPSH, CIAP could find some funding for its scientific projects.

Nordic researchers had eagerly supported international scientific cooperation in the interwar years (Rogan 2008a and 2008b), but they had kept away from CIAP, and they looked with scepticism at the 1947 circus in Paris.[4] Sigurd Erixon, the foremost actor within European ethnology and the one who had given the discipline its name in the 1930s, talked about the 'dilletantism' of the organization.[5] In spite of this, he made Sweden join CIAP in 1949, finding the prospect of UNESCO subventions attractive

2 For a detailed presentation of the genesis and development of CIAP in its early phase, see Rogan (2007).

3 Minutes/Compte-rendu sommaire des travaux de la 1ère session plénière (Paris, Musée de l'Homme, 1–5 Oct. 1947, MNATP: Org. APP-CIAP 1947–48–49 etc).

4 Letters from Campbell to Séamus Ó Duilearga and Seán O'Sullivan, 23.10, and 19.12.1947 (DUBLIN). Åke Campbell (Uppsala) has rendered his impressions from the Paris conference in letters to Irish colleagues, where he deplores the naivety of all the well intentioned proposals and decisions. I am grateful to Professor Nils Arvid Bringéus for drawing my attention to these letters.

5 Manuscript to Erixon's speech, 14.12.1949, on the occasion of the establishment of the Swedish national committee (SE).

Figure 2.1 Sigurd Erixon in the garden, probably from the 1960s.

Photographer unknown ©Nordiska museet.

and seeing a potential in the fact that the organization now covered the whole field of European ethnology – in principle if not yet in practice, and not only folk art and folklore, as had been the case before the war. The CIPSH-supported projects were first the bibliography, and later the dictionary of ethnological terms, the journal *Laos* and some 'specialist meetings', mainly on cartography.

The driving force in the scholarly work of CIAP in the 1950s was Sigurd Erixon (1888–1968), professor of ethnology in Stockholm and research director of the

Nordiska Museet. Erixon was in charge the dictionary of ethnological terms and he presided over what was probably the most active commission – on cartography. The most tangible result in other fields was the bibliography, organised by the Swiss *Volkskundler* Paul Geiger and Robert Wildhaber.

It should be observed, however, that the CIAP commissions were small and very different from the later SIEF commissions. They consisted of three to ten specialists, appointed by the Board. Only three commissions – the ones for cartography, the dictionary and the bibliography – were Reported to the CIAP Board. The latter two worked for a concrete output – the annual or biannual (and always strongly delayed) bibliographies, and the (likewise delayed) dictionaries.

The *Volkskundliche Bibliographie* had been a Swiss-German project since 1917, with Swiss editors (E. Hoffmann-Krayer, Paul Geiger) and German or Swiss publishers. In 1949, CIAP took over the responsibility through its bibliography commission, and the term 'international' was added; so it became *die Internationale Volkskundliche Bibliographie* (IVB). Robert Wildhaber (Basel), who remained its editor until 1977, built up a wide network of national referees (Wildhaber 1955, Brednich 1977/78). The dictionary of ethnological terms was originally proposed by Arnold van Gennep, but the work was carried out by Åke Hultkranz (Uppsala) and Laurits Bødker (Copenhagen). The dictionary commission under Erixon functioned as the editorial board for the two volumes published in the 1960s (Hultkrantz 1960, Bødker 1966).

The cartography commission worked for the coordination of national projects and practices, through a homogenization of techniques and methods, and common questionnaires and topics, and with a European atlas of popular culture as a distant goal (*Laos*, Vol. III 1955). The other commissions of CIAP tended to live their own lives, and their activities are difficult to trace, as they neither received support nor Reported back.

Erixon also edited the journal *Laos*, until UNESCO stopped supporting it in 1955. In Vol. I (1951) he presented his visions for the discipline. He saw regional ethnology as 'a branch of general ethnology, applied to civilised peoples, their social grouping and their complex cultural conditions'. Having abandoned his pre-war behaviourist and functionalist ideas, Erixon now found his inspiration in American cultural anthropology or 'culturology', with its concepts of culture areas, folk culture versus mobile culture, culture centres and ways of diffusion, acculturation and assimilation. His references are first and foremost American cultural anthropologists. He proposes a historical and comparative study of a field that embraces urban and industrial societies, societies in transformation as well as traditional societies. He advocates a study of culture in its three dimensions – space, time and social strata. The theoretical apparatus is that of diffusionism, and cartography is the tool *par excellence*. Such was the scientific programme that he recommended for *Laos*, for CIAP and for European (regional) ethnology – and which he also advocated through his cartography commission (Erixon 1951).

On the organizational side, however, CIAP struggled hard. The strict administrative regime before the war had been replaced in 1947 by a lax management. Between 1947 and 1953, the Board and the General Assembly convened only once – when Erixon was invited to *The International Congress of European and Western Ethnology* in

Stockholm in 1951, he sought to encourage discussions on international cooperation, yet the debate was inconclusive. When the Spanish CIAP president resigned shortly after, CIAP was thrown into a new crisis.

Several factors contributed to the crisis: a legitimacy dispute in the presidency, criticism from UNESCO for bad administrative management, missing archives, and delays in the publication of the programmes. Disorder in the accountancy (project subventions)[6] was discovered and CIAP was threatened with examination by audit experts.[7] The general secretary was forced to resign for not following up the decisions of the Board[8], for disorder in the finances and suspicion of embezzlement.[9] Sigurd Erixon and Georges Henri Rivière, the leader of *le Musée des Arts et Traditions Populaires* in Paris managed to keep above the quarrels in the secretariat. During the conferences in Namur (1953) and Paris (1954), they managed to restore order and to obtain permission to and reorganize CIAP.

Behind the crisis two interrelated major problems may be discerned: the resource situation and the membership structure. Pre-war CIAP had been based on national committees of folklore, as required by the League of Nations. The new CIAP of 1947 was based on individual membership. However, if it had been difficult to keep contact with the national committees, it was even more difficult with individual members. As the collecting of fees did not work, CIAP had no resources except the limited and earmarked project subventions from UNESCO. But UNESCO also required a membership of officially appointed national delegates, so Erixon and Rivière reintroduced the system in the second remoulding of CIAP. From 1954, CIAP was again based on national committees, which should appoint up to three members each to the General Assembly.

Erixon was urged to stand for the presidency, but decided to launch another candidate, the folklore professor Reidar Th. Christiansen from Oslo, who was elected and would remain the president of CIAP for the next ten years. Jorge Dias, Professor of ethnology in Lisbon, was elected as general secretary while Ernest Baumann, a folklorist from Basel, was elected to the position of treasurer.

The Paris meeting inaugurated a promising – albeit short – period in the history of CIAP. At the Arnhem congress in September 1955 and the follow-up symposium in Amsterdam, optimism still reigned. The scholarly focus was on cartography and especially on a European atlas for popular culture, and there was an important debate on the profile and scope of European ethnology and the naming of the discipline, especially whether the term 'folklore' should be used. The symposium in Amsterdam in 1955 – where, admittedly, ardent defenders of folklore as an independent science, like the Belgian Deputy President Albert Marinus – were not invited, stated in a

6 Letter of September 4, 1950, from J. Thomas, director of the Department for cultural activities (UNESCO. MNATP).

7 Note de Monsieur Marinus relative au Secrétaire-Général Monsieur Foundoukidis, undated/August–September 1953 (MNATP).

8 See stenographied minutes from the Board sessions, Namur (MNATP).

9 See correspondence (UNESCO, ICFAF Reg. 39 A01).

recommendation[10] that 'folklore' belonged to the national level, in countries where there was a tradition for this designation, whereas the name on the international level should be 'ethnology', for a field embracing spiritual, material and social culture. If desired, the qualifying epithet 'regional' or 'national' might be used, to distinguish it from (social) anthropology or the study of primitive or non-literate cultures. This consensus should not last long, however.

But problems on the administrative side should soon pile up again. The general secretary Jorge Dias (1907–1973) found the office strenuous, the membership system difficult to administer and the fees hard to collect. When the treasurer Baumann died, Dias refused to function as a treasurer, with reference to the fate of his predecessor. As no one was willing to take over the treasury, the economy of CIAP was a permanent headache. [11]

On the scholarly side, Dias experienced as an obstacle to CIAP and to the scientific development of the discipline what he defined 'an excessive love for the local and the particular' among the practitioners of folk life studies.[12] He also complained of the opposition from many folklorists to cooperate with ethnologists.[13] Dias' own position, that European regional ethnology was part of general anthropology, he shared with Erixon and Rivière, but it was not a widespread attitude in CIAP circles.

President Reidar Th. Christiansen (1886–1971) was an acknowledged folklorist scholar with a long experience from his international relations. But he was a prudent person who shunned conflicts[14] – and conflicts were precisely what he met in CIAP. He had long absences, when his research and periods as visiting professor led him to England, to Ireland and to the United States. During the first period of his presidency he spent two years abroad (1956–58), when he probably paid too little attention to CIAP. Furthermore, he turned seventy in 1958, his health was not strong, and as a retired professor he had no infrastructure to lean upon. The physical distance from Oslo to Paris, the seat of CIAP's only benefactor UNESCO, was also a complicating factor. When Jorge Dias resigned in spring 1957, Christiansen lost almost all administrative support and was stuck in a trap he never got out of. As he wrote to Sigurd Erixon in 1959, when the problems piled up: 'I regret sincerely that I did not resign, I too, when the secretary left. But I thought for honour's sake that I had to try and keep things going.'[15]

In July 1957, Winand Roukens, director of *Het Nederlands Openlucht-museum* in Arnhem, accepted the double function of general secretary and treasurer, after Dias and Baumann. But the inherent problems of the national committees and his

10 Recommendation from the Amsterdam meeting, September 1955 (MEERTENS 35:1131). See also Erixon (1955–56).

11 As for the archives, Dias seems to have kept them in good order, with the one curious exception that they suddenly disappeared in 1957, probably stolen. See letter of 7.1.1957 from Vieiga d'Oliveira to Dias (DIAS Box 4. Dias 4).

12 Letter of resignation from Dias to Christiansen, 16.4.1957; *Dias' Rapport moral sur les activités du secrétariat de la CIAP* (MNATP).

13 Letter of 4.5.1955 from Dias to Rivière (MNATP).

14 Interview with his daughter Elin Christiansen Smit, September 2007.

15 Letter of 3.2.1959 from Christiansen to Erixon (SE 8:28).

lack of success in collecting overdue fees, made him resign after only five months.[16] Roukens Reported that when he took over in 1957, there had been no fees collected since 1949, and he managed to get money in from less than a fifth of the registered members.[17] Between 1958 and 1961, CIAP was mostly without a secretariat.[18]

The bylaw of CIAP prescribed a Board meeting once a year, but between 1955 and 1964 there were only 2 regular Board meetings: Paris 1957 and Kiel 1959. In addition there were two 'expanded' meetings: Brussels 1962 and Bonn 1964. For lack of money to cover travel costs, all meetings were held in conjunction with folklore congresses. There was no General Assembly after 1955 nor did any congress take place. And there were no regular elections between 1954 and 1964. As UNESCO required that its member organizations hold regular elections and assemblies, the danger of exclusion was imminent.

These problems should by no means be attributed to the president only. With the exception of Erixon, and to some extent Marinus, the CIAP Board members were very passive. In a series of letters to Erixon, Christiansen repeatedly mentions the difficulties of getting response and support from the others for arranging meetings, and the more or less non-existent national committees seldom answered summons or invoices.[19]

CIAP's relations to CIPSH and UNESCO should become a recurrent theme in later discussions. From 1957 onwards, UNESCO repeatedly complained about lacking payments, as a small percentage of the membership fees should be returned to CIPSH. UNESCO also threatened to reduce or withhold the subventions for the bibliography and the dictionary, and periodically did it, as the publishing was strongly delayed. In 1957, UNESCO had signaled that a fusion with another member organization, IUAES (*The International Union of Anthropological and Ethnological Sciences*), was desirable, and in 1959 CIAP had to accept a *de facto* joint representation with IUAES in CIPSH.

By the end of the 1950s, CIAP was by most standards bankrupt and paralyzed. The president constantly sought advice from Erixon, his only confident, but he seemed incapable of taking any initiatives, nor of resigning, only hoping for more generous credits or asking for deferments from an unwilling and critical UNESCO administration. In this situation, Sigurd Erixon, ordinary Board member for all practical purposes, and with the consent of Christiansen, took over the leadership in CIAP, while Christiansen remained president in the name.[20] However, this 'Nordic

---

16 On the subject of collecting the fees, see letter of 16.10.1957 from Léopold Schmidt to Roukens (VIENNA CIAP Box 02). Roukens' letter of resignation of 8.1.1959 to Christiansen (ibid.; also in SE 8:28). It should be added that Roukens encountered some serious problems at his museum and was forced to resign from his post as museum director the same autumn.

17 Minutes from the Board meeting in Kiel, 18.8.1959 (MNATP, Org. App. CIAP).

18 In 1958, Sigurd Erixon persuaded the Swedish doctor Anna-Maja Nylén to function as secretary for CIAP, but she too resigned after a short time. See letter of resignation from Nylén to Erixon of 18.8.1959, containing some very critical remarks about the lack of response from Christiansen (SE 8:28).

19 See for instance the comprehensive correspondence between Christiansen and Erixon on the problems of arranging a Board meeting in 1958–59 (SE 8:28).

20 See correspondence between Christiansen and Erixon (SE 8:28, 8:30, 8:31).

alliance' was challenged by a much younger colleague, Kurt Ranke, professor of *Volkskunde* in Göttingen since 1960.

The Voksenåsen Meeting and Ranke's Memorandum

By the time of the Board meeting in Kiel, in August 1959, it was clear to everyone that the situation had become precarious, and Christiansen was invited to a meeting in Oslo to discuss 'the future organisation and activities of the CIAP'.[21] The meeting took place at Voksenåsen, in the woodland surroundings of Oslo, in late August 1961. The 75-year-old Christiansen had strived hard to find money to arrange 'this damned meeting', as he repeatedly calls it in letters to Erixon – who, for his part, could hardly conceal his impatience with the slow progress of the planning. Invitations had been sent to a considerable number of CIAP members and to some others. For once, money was available; still only two Board members turned up, Sigurd Erixon and Winand Roukens, plus three other foreign guests: Kurt Ranke, Paul de Keyser (Ghent), and Åke Hultkranz (Stockholm).[22] A few researchers from the Norwegian open-air museum filled up the meeting. Hans Nettelbladt was also present – a Swedish relative of Erixon[23] who had been requested to take on the treasury of CIAP. During the meeting Nettelbladt accepted to function as general secretary until the next General Assembly.

There are no formal minutes from the meeting, only a delayed Report written by Christiansen in May 1962 – after several impatient reminders from Erixon.[24] According to this Report, Christiansen had presented the state of affairs in CIAP, with a focus on the relation to the IUAES, there had been the usual Reporting on the current CIAP projects (the dictionary, the bibliography and the atlas work) and Hans Nettelbladt gave a statement on the financial situation. The Report ends with the stoic phrase: 'For the rest the financial problem is unsolved'. It appears that Kurt Ranke dominated the debate, coming up with several ideas and proposals for strengthening the economy of CIAP.

Just before Christiansen's delayed Report appeared, another text was circulated – entitled *Memorandum der Kommission zur Reorganisation der CIAP*, but called *Ranke's Memorandum* in later documents. In his own Report, Christiansen conceals the nomination of a committee at Voksenåsen, but according to Erixon, four persons – Ranke, Roukens, Erixon and Hultkrantz – had been asked to propose 'a new

21 Minutes from the Kiel meeting (SE 8:28).

22 de Keyser had been asked by Board member Albert Marinus to go in his place, and Åke Hultkranz, the author of the dictionary of ethnological terms, had been talked into coming by Erixon. Correspondence between Christiansen, Erixon and Hultkranz, letters of 8, 10 and 15.8.1960 (SE 8:30).

23 Hans Nettelbladt was a doctor of dental surgery from Stockholm and a nephew of Sigurd Erixon, and an amateur of folklore and ethnology.

24 The meeting at Voksenåsen, 27–29 August 1961. Memorandum by R. Th. Christiansen, dated May 1962 (MNATP Also: SE). The extreme delay of the Report was partly due to Christiansen's hospitalization and re-convalescence absence.

Figure 2.1 European conference on folklore, Brussels, September 1962

From the left: Sigurd Erixon (Sweden), Paul de Keyser (Belgium), Roger Lecotté (France), unknown person, Olav Bø (Norway). The photo is slightly reduced. Photographer unknown ©Nordiska museet. G28788.

programme [*Prinzipprogram*] and a new bylaw for CIAP'.[25] Ranke's draft, spurred by an impatient Erixon,[26] appeared in April 1962. Erixon welcomed the draft as a basis for discussions in the committee. President Christiansen, however, got so upset about certain elements of the text that he did not want the committee to continue, a committee which had not convened yet. Christiansen feared that Ranke's proposal to replace a commission (CIAP) with national representation by a society based on individual membership would mean an end to the UNESCO funding. The aged professor of folklore was also uncomfortable with Ranke's conception of the field as '*antropologische Wissenschaft*' and the idea of widening the scope of CIAP to become a *Verein für moderne kritisch-empirischer Social- und Kulturwissenschaft*. His formal argument was that this would make CIAP so similar to IUAES that it would lose its legitimacy in the UNESCO system. Christiansen also seems to have disliked the idea of a strong German dominance in CIAP.[27]

If Christiansen wanted to change as little as possible, Erixon was of another opinion. He saw much more clearly the necessity of reorganizing CIAP, and Christiansen could not prevent the spreading of Ranke's memorandum. From the summer of 1962, Erixon, with the help of the Swedish interim secretary, took on an active role and started planning the next Board meeting. With no financial basis for summoning a meeting, the first possibility for CIAP to convene was in Belgium in September 1962, where two consecutive events were being planned: a meeting in

25 Letters of 5.4.1962 from Erixon to Ranke (SE 8:27) and of 2.5.1962 from Erixon to Christiansen (SE 8:30).

26 Letter of 16.4.1962 from Ranke to Erixon (SE 8:27).

27 Correspondence (SE 8:27, SE 8:30, SE 8:31). See especially letters from Christiansen to Erixon of 29.4.1962 (SE 8:27) and of 5.5.1962 (SE 8:31).

Antwerp by Ranke and the *Fabula* circle, including the official foundation of the International Society for Folk Narrative Research[28] – an organization which Ranke presided for more than 10 years, and directly afterwards a *Volkskunde* congress in Brussels, to celebrate the 25th anniversary of the Royal Belgian Commission on Folklore. On Ranke's proposal Erixon took the initiative – with the consent of Christiansen – to a CIAP Board meeting in Brussels.

The Brussels Meeting and 'the Gang of Four'

Only four of eleven CIAP Board members attended the meeting: Erixon, Marinus, Roukens and Stith Thompson. Christiansen had not yet recovered from a complicated leg fracture, and no one took care of the secretarial functions. To Erixon's surprise, the Belgian organisers[29] convened the meeting on a day other than that agreed – in the afternoon directly after the closure of the congress, and not at the promised venue – but in a busy and noisy hotel lobby. Instead of convening only the Board and a few guests, as planned, they opened the doors to the participants of the congress. The same morning they had invited the congress in plenary to pronounce the following resolution: 'Having observed that there is not, for the time being, any international organization working actively to the benefit of folklore, the congress expresses the wish that this shortcoming be amended without delay'.[30] Or as Erixon dryly remarks in a letter to Christiansen: 'This implies a suggestion that CIAP is dormant'.

There are two Reports from the meeting. The one, which is unsigned but came to be claimed as the official minutes, is edited by the new committee of four (see below), on the basis of notes taken by the Belgian folklorist Roger Pinon[31]. The minutes were never sent to the participants for approval, and Erixon repeatedly claims in letters that parts of it are wrong.

The other one is a long letter from Erixon to Christiansen, written upon his return to Stockholm.[32] Erixon questions the legitimacy of letting members and non-members alike – without the right to vote – participate in the decisions; he had argued for following formal procedures and for contacting UNESCO before decisions were taken, but he had been overruled. The group of members and non-members had decided that a new reorganizing committee be appointed and proceed immediately to the task. Erixon gives – in his sober style, full of understatements

28 The decision to establish the International Society for Folk Narrative Research was taken at the congress in Kiel in August 1959, when CIAP was invited to arrange its Board meeting.

29 Albert Marinus and Paul de Keyser.

30 *'Constatant qu'il n'y a, en ce moment, aucun organisme international actif se préoccupant des intérêts du folklore, émet le vœu qu'il soit remédié sans retard à cette carence.' Réunion du Bureau élargi de la CIAP tenue à Bruxelles le 14 septembre 1962* (MNATP, Org. div. – CIAP 1962, 1963, 1964).

31 *Reunion du bureau élargi de la CIAP tenue à Bruxelles le 14 septembre 1962* (MNATP, Org. div. – CIAP 1962, 1963, 1964).

32 Letter of 12.10.1962 from Erixon to Christiansen (SE 8:31). Erixon explains that he wrote down what happened as there was no secretary present and no minutes to be expected.

– a description of a rather chaotic meeting, where Ranke and Wildhaber (not CIAP-members) dominated. Erixon's letter do not differ from Pinon's 'minutes' concerning Ranke's role; Ranke proposed both *who* should be members of the committee and *what* they should do, insisting that everyone loyally support the new committee. Erixon cannot refrain from mentioning that CIAP must have been vividly discussed at the preceding Antwerp seminar on folk narrative research. Four folklorists were elected as committee members–: Karel C. Peeters (Antwerp/Leuwen), Roger Pinon (Liège), Roger Lecotté (Paris) and Robert Wildhaber (Basel). Erixon's laconic final comment to Christiansen:

> The one-sidedness of the composition of the Board [meeting] may perhaps explain the stubbornness. Now the gentlemen in Belgium and France have taken the lead in connivance with Wildhaber and Ranke. Let us hope it will result in renewed vitality for CIAP. I have given a realistic description [... and not an official Report], so that you may have a clear understanding of the situation'.[33]

Of these four, only Robert Wildhaber (1902–1982) had some prior knowledge of CIAP, as the editor of the *Internationale Volkskundliche Bibliographie* since 1950. Wildhaber was the director of the *Schweitzerische Museum für Volkskunde* and a *Volkskundler* working with both folkloristic and folk art topics. He was the only one of the four with a broad international scholarly reputation. The other three were specialists in folklore (in the restricted meaning). Karel Constant Peeters (1903–1975) had recently been appointed to a chair in folklore at the private Catholic University of Leuwen. Roger Lecotté worked as a librarian at the National Library in Paris and Roger Pinon as a school teacher.

The committee, often referred to as 'the Four' – and who in their internal correspondence used nicknames like '*la bande des quatre*' or '*les quatre mousquetaires*' – held their first meeting in Antwerp in early November 1962, with Peeters chairing. They have left several hundred letters and mementos from this work, which can be traced in detail, and which indeed offers surprising reading. They produced three 'Reports', but they never bothered to make an analysis of the recurrent problems in CIAP, and their texts show a total lack of knowledge of CIAP's earlier history.

'The Four' started by editing their own mandate – the abovementioned 'minutes' from the Brussels meeting, which they used to legitimate later actions and decisions. The winner writes the history, one might say. Instead of writing a policy document first, where goals and objectives for the new organization are discussed, 'The Four' proceeded directly to a revision of the bylaw. The only real novelty in their proposal was the paragraph on membership. Instead of membership based on national committees, they proposed that both individuals and institutions (scholarly societies, museums, libraries, archives, etc) could adhere. And they gave the revised document a remarkable title: '*Statuts provisoires*' – i.e. a temporary bylaw. Instead of sending a proposal for a new bylaw to CIAP's Board, they distributed the 'temporary' one to around four hundred researchers and institutions in and out of Europe, for comments.

33 Ibid.

The package also contained the 'minutes', and thus the 'official version' could be spread all over the world.

The minutes from their November meeting,[34] intended for internal use, shows that 'The Four' were occupied by the relation to IUAES and the anthropologists and by the name of the organisation. However, the item that is given most attention – and the most astonishing reading – is a proposal for the distribution of positions within a new CIAP. Some 30 persons are mentioned by name; Erixon, Christiansen and ten others are safely put on the shelf as honorary members, whereas seventeen people ('The Four' included) are designated to the Administrative Council. The Executive Board is made up of six persons – half of it by themselves. On their very first meeting, Peeters proposed himself as executive vice president; Pinon should become general secretary and Lecotté treasurer. Richard Dorson (USA) was appointed president, while Raul Cortazar (Buenos Aires) and Matthias Zender (Bonn) became vice presidents.

One might have thought that this listing was just a joke (actually, it must have been a merry meeting, as Pinon returned to Liège in Peeters' overcoat, as he somewhat ashamed reveals in a letter of excuse the following day[35]). However, Lecotté describes their meeting in the following manner, in a letter to Peeters:

> From our *mousquetairien* discussions I hold a strong impression of comfort, confidence and hope for a CIAP that will be reorganised through a real 'blood transfusion'. The key positions were distributed with a natural logic, in a commonsense way and with a fortunate simplicity that augurs well for the future. I feel especially confident that the Belgian 'tandem' will contribute a straightforwardness and an easiness hitherto unknown to the Board member relations.[36]

The 'natural logic' is revealed a few days later in a memento, where Lecotté explains to the three others that he had written to Cortazar shortly after the meeting and offered to launch him as candidate for the position of vice president. In 1960 Raul Cortazar had founded an international South American association, *Comision Internacional Permanente de Folklore* – CIPF. Believing that CIPH was now dormant, 'the Four' wanted to secure the support of Cortazar and thus obtain a great number of South American adherents. To Lecotté's surprise, Cortazar answered that CIPF was by then a vital organization, even applying for UNESCO/CIPSH membership. He politely refused the offer, but welcomed cooperation. An alarmed Lecotté Reported to the three others that they had a new competitor for the UNESCO funds, instead of an accomplice.[37] 'The Four' were not any luckier with the other vice president proposal. Matthias Zender refused the offer – with the addition that he felt 'really ashamed'

34 *Compte-rendu des travaux de la Commission de réorganisation de la CIAP en sa réunion des 1 et 2 novembre 1962 à Anvers* (PEETERS 1).

35 Letter of 4.11.1962 from Pinon to Peeters (PEETERS 4).

36 Letter of 4.11.1962 from Lecotté to Peeters (PEETERS 4).

37 '*Note pour les membres du Comité de réforme de la CIAP*' of 18.11.1962, letter of 18.11.1962 from Lecotté to Peeters of 18.11.1962 (PEETERS 4).

about their intentions.[38] But they cannot have been much 'ashamed' themselves; a few days later Pinon wrote to Peeters: 'As for the headquarters of CIAP – if the secretariat rests with me, we must find a place in Belgium ... I think it would be very nice to have it in the *Palais de Congrès* in Brussels, or in the *Cinquantennaire* [i.e. one of the most exclusive museum and park areas in Brussels]'.[39]

These actions can be thus explained: the committee regarded itself as the functioning Board of CIAP already, and it took for granted that the 'temporary' bylaw would be passed. According to the proposed procedural rules, it was the Board who should set up the election slate – i.e. propose candidates for the offices. Even before the temporary bylaw was sent out for comments, the committee had started practicing them. Erixon had had his misgivings by the end of the Brussels meeting, when he had explicitly exhorted from the participants a confirmation that they still regarded CIAP as active and existent, and that its elected officers and commissions should remain in function until the next General Assembly. This is left out of the 'official minutes' from Brussels, produced by 'The Four' themselves, but it is Reported in detail in Erixon's letter.

Also, in their communication with UNESCO and other international organizations, 'The Four' acted as if they were the Board of CIAP. Already in December 1962, they claimed to have '*pleins pouvoirs*' – full powers – to change the bylaw of CIAP before the end of 1962 and to begin immediately a process of rejuvenation of its officers.[40] During the winter of 1962–63, 'The Four' started negotiations with CIFP (above) and talks with UNESCO officials, and they continued to approach their selected candidates; in May Richard Dorson accepted to run for the presidency.[41] These activities were kept secret to the (old/legal) Board and several of their internal Reports and letters are marked 'Confidential', especially their Reports from the fairly regular contacts with the secretary of CIPSH, Jean d'Ormesson.[42] During the spring and summer of 1963 'The Four' started an intense campaigning, sending no less than around one thousand letters[43] to researchers and institutes all around the world, asking for their personal adherence to the 'new CIAP', and they managed to have around one hundred signatures before the end of the year.

At their meeting in May 1963, 'The Four' went the whole way, when discussing their legal position. The minutes[44] – intended for internal use only – state that: 'On the question of the authority of the Reform commission, it should be noted that the instructions given to the Commission by the Brussels meeting include the preparing and convening of the reconstitutive assembly. This implies the setting of the agenda and taking all necessary steps for a successful implementation. The agenda may have the following items'. According to Erixon's Report, as well as his reactions in

38 Letter of 8.11.1962 from Zender to Lecotté (PEETERS 4). On the other hand, Zender offered his collaboration.

39 Letter of 7.12.1962 from Pinon to Peters (PEETERS 4).

40 Commission [...] CIAP et CIPF. Exposé de la question. Archives (PEETERS 4).

41 Letter of 24.5.1963 from Wildhaber to Peeters. (PEETERS 4).

42 See for instance *Compte-rendu d'entretien entre M. J. d'Ormesson et Roger Lecotté au secretariat du CIPSH ... Ier février 1963* (PEETERS 4).

43 Letter from Peeters to Ranke of 4.6.1963 (PEETERS 4).

44 *Compte rendu de notre réunion du 11 mai à Paris* (PEETERS 1).

letters when this news reached him, no such directives had been given in Brussels;[45] to the contrary, the majority had confirmed that they regarded CIAP's Board and elected officers in function.

Words like 'putsch', 'rebellion' and 'revolution' were used by several observers, and the rumours reached UNESCO. Lecotté Reported – confidentially, as usual – from a meeting with the CIPSH secretary in November 1963: 'Mr d'O. also told me that some people think that our action is a rebellion against the old CIAP, that we are manoeuvring on the outside, impatient to take their places ... without respecting a regular transfer of powers ...'.[46] The legality of many of their actions and dispositions was a constant worry to 'The Four'; letters and documents produced by the committee repeatedly invoke the 'legal right', allegedly acquired at the Voksenåsen and the Brussels meetings – and there are several cases of scarce precision and creative history writing.[47]

From spring 1963 CIAP had in practice two competing executive organs. One was the legally elected (old) Board and Presidency. The 'legality problem' of the (old) Board was that they had by far exceeded their election period,[48] a fact that may have made it easier for many to accept the irregular procedure of 'The Four'. The other executive organ was 'The Four', who based their position on a mandate from non-members and a dubious document, the 'minutes' reconstructed by themselves.

In order to understand the following escalation, it should be observed that Christiansen, Erixon and Nettelbladt for a long time ignored that 'The Four' claimed full powers. When Erixon during the spring of 1963 continued his efforts on behalf of Christiansen to arrange a Board meeting, and when Nettelbladt sent invoices for the yearly CIAP fees, this normal activity was received by 'The Four' as provocations – or rather as pretexts to strengthen their propaganda by open letters and other actions.

'The Four' wanted '*les Nordiqeus*'[49] to know as little as possible of their planning and campaigning. This is clear from their notes marked 'Confidential' and from

45 The above-mentioned Report from Erixon to Christiansen, plus several other letters and memos (Nordiska museet, SE). See also letter from Peeters to Ranke of 27.12.1963 on the authority invoked by the 'the Four' (PEETERS 4).

46 *Note pour le comité des quatre (Confidentielle); Entrevue du 25.11.1963: Mr d'Ormesson – Roger Lecotté* (PEETERS 5).

47 Some documents, like the proposal to make CIAP an umbrella organisation for CIPF (South America) and other continental organisations (PEETERS 4) – written by Lecotté (undated/ Dec.1962) – after he discovered that CIPF was still going strong, see text – is a clear case of rewriting history in order to invoke legality. See also letters from 'the Four' to Rivière, d'Ormesson, etc., where they even claim that Erixon voted for their 'full powers' (Ibid).

48 The fact that no elections had been held for a long period was criticised by the CIPSH authorities. It was also discussed by Christiansen and d'Ormesson, the CIPSH secretary in Stockholm in May 1963, but CCIPSH accepted that the elections be Reported to the first possible General Assembly, that is in Athens 1964.

49 The word *les Nordiques* is normally used to signify people from all Northern European countries (which may be the case here, as there is question of the concept of folklore), but sometimes more specifically to signify the Scandinavians. When the 'the Four' talk about *les Scandinaves*, the meaning is clear, but when they use *les Nordiques* it is a question of context.

letters and memos where strategies are constantly discussed. They discuss how they should proceed in order to surprise their (supposed) opponents concerning the name of the organization, as in the minutes from their May 1963 meeting: 'it was decided to propose, in the Report to be sent out [no 2], that the new name should be *Conseil International de Folklore*, and not *Société* ... In order to prevent a manoeuvre from '*les Nordiques*', we may propose to continue to use 'CIAP', and then we may come back later with the new designation'.[50]

Or in a letter from Peeters to Lecotté in June: 'I agree with your remarks on the terminology. We will talk it over confidentially next time we meet and officially at the next plenary, when the Scandinavians and other technologists are present'[51]. Their strategies could go very far, as when Lecotté Reports from one of his several lobby meetings with the secretary of CIPSH:

> *Confidential*: Jean d'O. [d'Ormesson] has revealed to me that Christiansen has written to CIPSH to ask for subventions. *These are already available*. Jean d'O. asked me what to do? – I answered that if Christiansen were granted these funds he would immediately gain in importance in relation to us ['the Four'] and our efforts would have had no other effect than to maintain his position. J. d'O. agreed ... and as the application for money will be treated at the CIPSH congress, he will propose that the money be granted to CIAP but *withheld until the reorganization and the new elections are carried through*. J. d'O. believes this will be accepted by the CIPSH, when they are informed about our work ... and that they will dismiss Christiansen's request.[52]

The same Lecotté knocked regularly on Rivière's door in Paris, in order to convince him to join their cause, to abolish the system of national committees and to accept the word 'folklore' in the title of the new organization. It seems clear that Rivière disliked these visits, both because he disagreed on several points and because he was in serious doubt about the legitimacy of the actions of 'The Four'. The quotation below, from one of Lecottés 'confidential Reports', reveals some of the things at stake for 'The Four':

> He [Rivière] will absolutely not use the word *folklore*, he will impose the word *ethnology*, as this was decided at the Nordic congress, as he says ... He hardly acknowledges the popular tales and beliefs, but seeks refuge in the technology, like Erixon. We are far away from the true folklore ... And all this only to be accepted by the gentlemen at the University. In fact (and this between us) he has been drawn in this direction by his great friend Cl. Levy-Strauss, whom I consider a very intelligent person, but dangerous in the sense that he deconstructs everything and constructs nothing. His speeches and his works give immediately a brilliant impression. But after some reflection they leave behind

50 *Compte rendu de notre réunion du 11 mai à Paris* (PEETERS 1).

51 Letter of 28.6.1963 from Peeters to Lecotté (PEETERS 4).

52 *Note urgente pour Karel Peeters et Roger Pinon. Entretien du mardi 13 août 1963: Ormesson – Lecotté* (PEETERS 4). One sometimes wonders whether the notes from Lecotté and Pinon refer to real events or how much fiction there is in it. The otherwise highly respected Jean d'Ormesson – French author, journalist, civil servant, editor-in-chief, etc – gave very different information to Christiansen and Erixon, cfr. letters from Christiansen to Ranke, 12.5.1963 and to Lecotté 13.5.1963 (SE 8:31).

> a feeling of emptiness and chaos; his principle is to make everything dissolve (that's typical for his race, by the way). We must rally a majority to combat these technologists (technocrats) as effectively as we can.[53]

If we disregard the anti-Semitic aspect, two elements in this quotation contribute to a deeper understanding of the aggressive campaigning of 'The Four': the fight for an immaculate folklore, and the striving of folklore (in a restricted sense) to gain recognition as an academic discipline in some countries. 'Technologists' or 'technocrats' were expressions used by Lecotté, Pinon and Peeters for ethnologists who studied material culture.

We shall return to some of these questions later. The epistolary sources disclose the strategies of 'The Four' – but also the absence of strategies in the other camp. Erixon learned for the first time that 'The Four' had claimed full powers when he received in mid-June 1963 the first open letter, addressed to him. He wrote to Christiansen:

> Received today the enclosed protest letter from the revolution committee in Antwerp. I call it so, as it claims all powers. To me it is strange reading. I will not comment on the legal side of it. In the minutes from Brussels (which I have never fully accepted in some details) I have not found any basis for these claims, but I know that some people talked in a manner that sounded revolutionary… For my part, it is easy to answer, as they have misunderstood both my intentions and my decisions, not to speak of how they have misinterpreted the word 'working group'. This is sheer nonsense.[54]

A few days later Erixon sends a note to the secretary Nettelbladt:

> I have understood that you did not have time to bother about CIAP. But the train moves on. Yesterday I got a message from Christiansen who informed me that he had received a copy of the [open] letter to me from 'the revolutionary committee', but he thought I should ignore most of it. This evening however the main grenade hit the president's stronghold, through a complicated document (a very hard personal attack on Christiansen)[55].

Little did Erixon know that a much heavier 'gun' was just being loaded and aimed at Oslo – this time from Göttingen. Shortly after the impact, that is the publishing of a new open letter, from Ranke (see below), Erixon stoically wrote another note: 'I have consulted Hans Nettelbladt the 12/9. We decided to await some sort of sign of life from Christiansen. As things have moved now, it is best to stay neutral. We have tried to protect Ch[ristiansen], but I think it's all over with him now'.[56]

The reading of many of these documents is so surprising, and the campaign of 'The Four' so aggressive, that one may wonder what was at stake, beyond the apparent territorial fights. Why did 'The Four' distribute their Reports to around four hundred scholars and institutes, why send out around one thousand pre-printed formulas for adherence, and why attack 'the establishment' through open letters,

---

53 Letter from Lecotté to Peeters of 24.6.1963 (PEETERS 4).

54 Letter of 17.6.1963, Erixon to Christiansen (SE 8:31).

55 Letter from Erixon to Hasselbladt of 25.6.1963 (SE 8:31).

56 Note by Erixon of 12.9.1963 (SE 8:31).

spread to a broad scientific community? An important explanation is probably to be found in the structural constraints. CIAP was very difficult to change from within. Its bylaw prescribed voting rights for few persons – as its General Assembly was composed of up to three delegates from each national committee. And the delegates had a right to vote only if the fees had been paid – which was the exception rather than the rule. These rules were laid down by UNESCO, and radical changes in the bylaw implied the risk of loosing the seat in CIPSH and consequently the UNESCO funding. A revolution from the inside of CIAP was almost as difficult to imagine as a rising of the buried on a churchyard; hence the usurpation of full powers.

Christiansen's dream was to strengthen the national committees, to make them function as intended. And when the secretary renewed his efforts, in spring 1963, to collect the overdue fees, it *could* be understood as a countermove to make the old system function; hence the strong reactions from 'The Four'. Their logic was that the system could be changed only from the outside, as long as the votes were reserved for the old establishment. They had to instigate a popular movement – through hundreds of postal missives and the spreading of minutes (whether forged or not), Reports and open attack letters. Also, they had to conquer the right to convene the General Assembly – which could not be a traditional one, but an 'expanded' or a broad and open-to-all assembly, with voting rights to all present. 'The Four' claimed to have a democratic approach; Erixon called it revolutionary. Erixon claimed to have a parliamentary approach; 'The Four' called it despotic: We cannot here follow in detail the work of 'The Four', which lasted almost two years and has left a rich heritage of letters and documents. In the following paragraphs I will give a survey of the events and focus on some main actors and some documents.

On the Road to Athens

After the Brussels meeting (September 1962), there was no regular CIAP reunion before the General Assembly in Athens in September 1964, held in conjunction with a folk tale congress arranged by Ranke's International Society for Folk Narrative Research. In April 1964 the CIAP question was discussed in Bonn at a meeting planned by Kurt Ranke, during an atlas conference hosted by Matthias Zender.

As noted, 'The Four' met in Antwerp in November 1962, in Paris in May, September and December 1963, and in Bonn in March 1964. In addition, the two Belgians – Peeters and Pinon – saw each other regularly, and there was an intense correspondence, especially between Peeters, Pinon and Lecotté. Wildhaber passed the academic year 1963–64 in USA, invited by Dorson to Bloomington, and for periods he was less active than the three others in the committee, but judging from the correspondence, he kept regular and intimate contact with Ranke (letters which reveal the sharp and blunt characters of both adversaries and allies).

'The Four' wrote a Report on the basis of the thirty to forty incoming comments to the first *Statuts provisoires*. Report no. 2 was distributed in late May 1963 together with a revised proposal for bylaw, for a new round of comments. This came as a surprise to Erixon and Christiansen, who had thought that the committee would complete its mission by a Report to the Board. Early in 1964, Report no. 3 appeared, which was based on about one hundred incoming letters as well as a series of minor gatherings, bilateral meetings, sessions at folktale conferences, etc. The congresses in Santo Tirso (Portugal, June 1963) and Budapest (October 1963) especially served as fora for discussions on CIAP. Reports no. 2 and 3 are in greater part worked out by Pinon, the intended general secretary of the 'new CIAP'.[57] Report no. 3, with an adjusted bylaw, was the basis for discussions in Bonn in April 1964. During 1963 and 1964, CIAP seems to have been discussed in every corner of Europe, and also on some North and South American university campuses.

Whereas the first bylaw proposal had been more or less a copy of the bylaw from 1954, the May 1963 version is much more elaborate. A notable difference is also the change of title. The document is no longer called *Temporary bylaw ... by the Reorganizing committee*, but *Proposed bylaw for the adoption of the General Assembly*.[58] This did not mean, however, that 'the Four' did not still invoke the executive authority of CIAP. Another important difference is that 'arts et traditions' is replaced everywhere in the bylaw by the term 'folklore', except in the name of the organization (cf. the strategy revealed in the minutes). Also, Report no. 2 presents a set of arguments for using the term *folklore* as well as for adopting a more folkloristic profile.

The flow of comments to the Reports covers hundreds of details. Among the most important issues were the questions of national committees, of the disciplinary field(s) to be covered, of the definition of *folklore*, and of the name of the organization. Opposition to leaving the system of membership based on national committees came from Rivière, Erixon and Christiansen. Gerhard Heilfurth (Marburg), himself president of one of the few functioning national organizations – the *Deutsche Gesellschaft für Volkskunde*, also reacted negatively to only individual membership.[59] The argument for maintaining the national committee system, preferably in a strengthened form, was not only UNESCO's formal requirement. It was argued that national committees were indispensable for planning and organizing big international projects, like the European atlas. And there was the 'democracy' argument: designated national delegates would secure a more equal and even representation and prevent that special groups dominate the organization. But the majority wanted individual membership. To all those who would never have a chance of being designated, individual membership seemed more 'democratic'. By addressing the proposal to a large number of scholars, 'the Four' would logically get support for individual membership.

Not surprisingly, the proposed name of *folklore* for the discipline, and for the field(s) to be covered by a reorganised CIAP, was to many a provocation. Substantial parts of Report no. 3 (35 pages) deal with nomenclature and delimitation issues. It is far from clear what the committee means by folklore, in Pinon's long, intricate

57 *Note pour le comité des quatres* by Peeters is a detailed Report of several CIAP discussions in Budapest, 17–19.9.1963 (PEETERS 5).

58 *Statuts proposés à l'adoption par l'Assemblée Générale*. The statutes and the 2nd Report are found in several archives (MNATP, Org. Div. – CIAP 1962, 1963, 1964).

59 *Stellungsnahme zu den Vorschlägen der Reorganizationskommission der 'CIAP'*, 20.6.1963, signed Heilfurth and Weber-Kellermann (PEETERS 5; SE 8:31).

and not always coherent text – a fact that was addressed by several commentators. A complicating factor is that the debate was about the delimitation of the field(s), the name(s) of the discipline and the name of the organization.

Commentators, not least the Eastern Europeans, reacted to what might be read as a defence for the supremacy of folklore: some were very critic, like Wolfgang Steinitz (East Berlin); others like Branimir Bratanić and Milovan Gavazzi (both Zagreb) were moderately critic.[60] In this respect, the response of Prof. Axel Steensberg (Copenhagen), who had not taken part in any of the discussions, is exemplary and worth mentioning here:[61]

> Personally I feel that material culture is outside the French and Belgian colleagues' interests. And as I am a professor just in that subject, I do not wish personally to join the new organization, nor do [sic] my institute. But that is not to say that we are hostile against it ... The papers you had forwarded were very confusing according to the discussion. Not even the explanation of 'folklore' is sure, and the connection with [IUAES] seems to be rather vague.

When Steinitz, earlier vice president of IUAES, reminded of the meaning of the term *folklore* in Northern and Eastern Europe and asked for a plan for a division of labour with IUAES, 'the Four' retorted aggressively almost holding Steinitz up to ridicule. The result was a polemic debate, which ended with a turnabout from Peeters, in the form of a submissive letter, and full reconciliation.[62] 'The Four' realised that they needed allies in Eastern Europe, and Steinitz – one of the foremost *Volkskundler* behind the Iron Curtain, would have been a difficult opponent.

Report no. 3 maintains that *folklore* – which they wanted to use in the name of the new organization – covers both spiritual and material culture, but the text is still full of ambiguities, as are also the examples used. Many critics raised their voice against *folklore* in the name of the organization, whereas others expressed their satisfaction: 'I am glad to see that they are inclined to retain the name Folklore', wrote Stith Thompson.[63] Another difficulty was the constantly negative description of the field that is the fact that its limits *against* the disciplines of anthropology, ethnology and ethnography were highlighted (and reflected the committee's strong opposition to being merged with IUAES).

A main reason for the lack of clarity of the Reports is that 'the Four' themselves held different opinions in these questions. Pinon and Lecotté adhered to a restricted conception of *folklore* as spiritual culture. In their internal correspondence, the two repeatedly argue for the 'true' or the 'pure' folklore as a distinct scientific discipline, and they do not hide that they would prefer to make CIAP an organization for folklore in a restricted meaning. Peeters and Wildhaber, on the other hand, included both spiritual and material culture in the concept of *folklore*. Lecotté and Pinon insisted on *folklore* as the sole term in the name of the organization (a stand they hid as

60 Div. documents (PEETERS 3).

61 Letter of 26.6.63 from Steensberg to Wildhaber (PEETERS 5).

62 Div. correspondence (PEETERS 4, 5, 8).

63 Letter of 27.4.1964 from Thompson to Ranke (GÖTTINGEN).

long as possible, to avoid a confrontation with *les Nordiques*), whereas Peeters was indifferent to the name.

Report no. 3 contains a description of a future CIAP which was remarkably unrealistic, for its conception of the discipline as well as its organizational framing. The 'new CIAP' is presented as an 'international centre' for a 'synthesis of folklore', where 'the laws of folklore' should be unveiled. 'The Four' propose a worldwide organisation, based on continental sub-organizations (Africa, America, etc.), and with the European one hosting the central administration – hence perhaps Pinion's dream of the *Palais de Congrès* as the headquarters. And they proposed a hierarchical cycle of recurring congresses – on national, international and intercontinental levels – which should offer respectively local, comparative and synthesizing approaches.[64] Not less remarkable – in a committee Report – is the overt attack on Christiansen and Erixon, and the no less overt subjection to Ranke, with reference to his *Memorandum*.

The strength of the attack must be seen against the backdrop of an awkward incident – the cancellation of a CIAP Board meeting – between the writing of Reports 2 and 3. As this incident contributed in a decisive manner to a worsening of the climate, a brief rehearsal is necessary.

In the winter of 1962–63 Kurt Ranke – still as a non-member – took an initiative to a CIAP Board meeting in Göttingen in September 1963. Christiansen and Erixon, who were on the lookout for possible venues and funding for convening the Board, in order to conclude the process, were positive to his initiative and accepted the offer. However, an unexpected possibility that turned up in January 1963 gave hope for the funding of a Board meeting in Stockholm as early as June, as well as offering a venue for a General Assembly. The recently established Wenner-Gren Center in Stockholm and its Nordic Council for Anthropological Sciences offered to host a CIAP meeting, in conjunction with the council's own planned seminar on anthropological research in Europe and UNESCO's role. Christiansen agreed, and so did Ranke – who withdraw his proposal for Göttingen.

Erixon started planning a broad gathering, for both the Board and the General Assembly as well as for the former Voksenåsen and the present Brussels committees. And he followed Ranke's advice to make thorough preparations, by setting up a local organizing committee.[65] On a visit to UNESCO the same spring, however, the Nordic Council's secretary learned that CIAP was no longer considered a full member of CIPSH,[66] but only a sub-commission under IUAES. To the surprise of Erixon, CIAP had been «degraded» in the UNESCO system because its national committees had

---

64 It was Pinon who drafted the document, and Lecotté and Peeters who discussed it. Wildhaber did probably not participate in the production of the texts, at least not in no. 3, as he passed the time in Bloomington

65 *Ich würde jedoch unbedingt empfehlen, das Program dieser Sitzung sehr sorgfältig auszuarbeiten*, Letter of 26.1.1962 from Ranke to Erixon (SE 8:31).

66 In 1957, CIPSH was so dissatisfied with CIAP that was forced to share a seat with IUAES (the *International Union of Anthropological and Ethnological Sciences*), i.e. having one common delegate – in practice a member of IUAES – who defended the interests of both organizations at the yearly CIPSH meetings. Then again, in 1959, CIAP's formal status in the UNESCO system had been reduced to that of a branch of IUAES. However, its projects (the bibliography, etc.) were still supported economically.

not paid their fees. The result was that the Nordic council lost interest in CIAP, and after long deliberations, it offered a reduced contribution towards the travel costs for Board members only, for them to join the council's seminar in June. Erixon, who was still very intent on convening the Board in order to discuss a new bylaw and the UNESCO status, sent out invitations to the Board. By the end of March, however, the Nordic Council unexpectedly changed the date of their council to the beginning of May. With a far too small grant at his disposal, a changed schedule and too short time to organise the meeting properly, Erixon had to cancel the meeting.

In the letters of cancellation Erixon informed that Christiansen would meet with the CIPSH secretary in Stockholm, in order to gain information about CIAP's present status. He also informed that the local organizing committee would discuss the bylaw proposal, in order to prepare the next Board meeting.[67] Erixon's initial initiative to a Stockholm meeting was in itself badly received by 'the Four', but the information about talks with d'Ormesson and a meeting of a local organizing committee was possibly understood as a manoeuvre to bypass them. At least it provided a pretext for an escalation of the propaganda war.[68]

Shortly after the meeting, in mid June 1963, Christiansen wrote to Ranke and 'the Four'. He Reported that he had got green light from CIPSH, including allocations for 1963-64, acceptance to defer the convening of a General Assembly to September 1964 (Athens), as well as a special allocation for this event.[69] This meant that Erixon now controlled the train of events, in the eyes of Ranke and 'the Four', because the right to convene the Assembly and to define the agenda was of vital importance. When Christiansen also signalled, with reference to his talks with d'Ormesson, that the proposed bylaw changes could easily be adapted to the present bylaw and an improved system of national committees, and that the name of CIAP could be kept, the alarm bells must have rung. Lecotté formulated it this way:[70] 'Christiansen's letter gives food for thought. If we do not convene an Assembly, all we have done will be *neutralized*, to his advantage; he will take all the credit and in Athens people will be impressed by him.' And Peeters wrote to Ranke: 'It is clear from this letter that we must make an effort if we still want to secure CIAP.'[71]

The countermove was two letters from 'the Four', one to Erixon and one to Christiansen. The letter to Erixon contained some reproofs and inquisitorial questions, and it held an aggressive but not impolite form.[72] This letter was 'open', in

67 See correspondence (Erixon, Christiansen, Ranke, Marinus, Roukens) – in total 20–30 letters (SE 8:30-31).

68 'The Four' as well as Ranke were certainly informed about this misunderstanding, from Rivière, who had been present. The comment to Rivière's explanation, however, was: '[...] *dieses hat nichts an unseren CIAP-Plänen geändert.*' Letter of 4.6.1963 from Peeters to Ranke (PEETERS 4).

69 See the minutes from the Stockholm meeting (Christiansen, Erixon, Nettelbladt) 7.5.1963: *Protokoll fört vid sammanträde tisdag den 7 maj 1963 ...* (SE 8:31).

70 Letter of 12.6.1963 from Lecotté to Peeters (MNATP, PEETERS 4).

71 Letter of 4.6.1963 from Peeters to Ranke (MNATP, PEETERS 4).

72 Letter of 8.6.1963 from Lecotté, Peeters, Pinon and Wildhaber to Erixon (SE 8:31; PEETERS 4).

the sense that copies were distributed to around thirty people.[73] However, Erixon was not informed about this broad distribution, so he had no chance to correct publicly the misinformation it contained. The letter to Christiansen, 'the main grenade' as Erixon called it, was a frontal attack on both Christiansen and the national committee system, with the following conclusion: 'All we do is to catalyse the discontent and the discouragement of all the members, which are expressed on all levels by those who have been in contact with our organization. Your nine years of presidency is the most disappointing in the whole of CIAP's history. Why don't you draw the conclusion?'[74]

The activities of 'the Four' would have had little or no interest to posterity, had it not been for the fact that it was they who managed to rally the scientific community and to conquer CIAP. But whose work was it? My assertion is that their plans, which on several points lacked a sense of reality, and their propaganda, which is best characterised by its bluntness, would never have succeeded – without a better strategist in the background.

A Grey Eminence and his Henchmen

It is commonly held by posterity that Kurt Ranke, professor in Kiel from 1958 and in Göttingen from 1960, had no special relations to CIAP and SIEF.[75] Except for his election to the SIEF Council in 1964 (as one of fifteen members), it is correct that he never held any office in these organizations, but his role was no lesser for that.

Ranke (1908–1985) was an ambitious and enterprising scholar, who by the end of the 1950s was regarded as one of the foremost folk narrative researchers. After World War II he was among the first German *Volkskundler* to establish a broad net of international contacts, including Nordic as well as American folklorists. The Scandinavian folklorists observed him with some apprehension in the late 1950s, as can be seen in a letter to Christiansen from Laurits Bødker – the author of the second volume of CIAP's *Dictionary of Ethnological Terms* – in the summer of 1959:

> There is one problem, however, that I think we should observe that is whether Ranke is working to establish a new, international *Märchen* organization. I don't know if this is the case, but some of his remarks during the last year indicate that such are his plans. In that case, we need to know beforehand approximately what he is aiming at. Will he establish

---

73 See letters of 4.6. and 13.6.1963 from Peeters to Ranke and Lecotté (PEETERS 4). See also list of 29 addressees, ibid.

74 *Proposition de lettre à Christiansen, signée Lecotté, Pinon, Peeters and Wildhaber* (MNATP, PEETERS 8). The original letter has not been found, but according to Christiansen's answer it was dated 10.6.1963.

75 For a full assessment of the role of Kurt Ranke in the CIAP reform process, it would have been highly desirable to have access to his archives in the *Encyclopädie des Märchens*, under the *Akademie der Wissenschaften zu Göttingen*. However, for reasons that are not clear to me, I have been denied access to these archives, and it is not possible for me to judge whether these archives would throw further light on Ranke's involvement with CIAP and SIEF. The archives at the Georg-August-Universität Göttingen, where I was given access, contain only a small fraction of Ranke's correspondence on CIAP/SIEF.

an international centre like CIAP's secretariat of plowing implements in Copenhagen? What will be the relations between this new organization and CIAP? Will he establish *Fabula* [a scientific journal of narrative research started by Ranke in 1958] as a central organ, and will he launch a new monograph series? These questions are important for the Nordic countries, where we have already organs and series that we must protect, that is *Arv* and FFC [*Folklore Fellows Communications*].[76]

It is possible that apprehensions like these contributed to Christiansen's negative reactions to Ranke's *Memorandum*. The 'new, international *Märchen* organization', the ISFNR, was launched at the Kiel congress in 1959 and formally inaugurated in Antwerp in 1962. It was at the Kiel congress, where Ranke had offered a slot for the CIAP Board meeting, that he proposed to save CIAP's bibliography the *Internationale Volkskundliche Bibliographie* (IVB) from bankruptcy, through a considerable economic contribution from the German *Bundesministerium*.[77] The UNESCO subventions and the sales were too low to keep the project going in Switzerland. The condition was that a German publisher took over the publishing – as had been the case before the war – and the responsibility be transferred to the national German *Volkskunde* association, but with Wildhaber still as editor. The solution actually secured the continuation of the IVB.

One interesting detail from the Kiel points forward to the Brussels meeting in 1962. When the vacant positions as treasurer and general secretary were discussed, Robert Wildhaber came up with two candidates: Roger Pinon and Roger Lecotté.[78] Three years later, the alliance Ranke–Wildhaber–Pinon–Lecotté should become a very decisive factor in the reform work.

Ranke's role at the next CIAP meeting, at Voksenåsen in 1961, has been presented above. Ranke once more proposed to assist a weak CIAP, by inviting to a close collaboration between CIAP and the German *Archiv für Volkserzählung*. The most important output of the Voksenåsen meeting, however, was *Ranke's Memorandum*, which should become one of the most decisive documents in the history of CIAP – not so much for its contents as for its role as a symbol and an incitement for turning the page. The title is *Memorandum der Kommission zur Reorganisation der CIAP*,[79] and it is both undated and unsigned. It presents itself as the work of a committee appointed in Oslo in 1961, but as we know, Ranke was the sole author.

Ranke's Memorandum (3–5 pages) opens with a declaration on how regional ethnology during 'the last decade' had developed from a romantic-mythological research on relics to a modern critical-empirical science covering culture and social life – an introduction that annoyed Scandinavian researchers. Erixon states that the postulated recent romantic-mythological approach might be valid for Germany, but

---

76 Letter from Bødker to Christiansen, 31.7.1959 (OSLO Ms 4 3516 Reidar Th. Christiansen IV Brev).

77 Minutes; *Réunion du bureau de la CIAP, Kiel*, [...] 1959 (MNATP: Org. App. CIAP. Dossier: CIAP 1955 à 1961). Ranke collaborated with Professor Helmut Dölker, Stuttgart, for this operation. Dölker had been the German member of the CIAP Board since 1954, but he seems to have taken no interest in CIAP affairs.

78 Ibid.

79 PEETERS 3. Also in French translation (MNATP, in SE, etc).

not elsewhere.[80] The document describes very briefly the contemporary challenges to the discipline of *Regionalethnologie* – comprising *Volkskunde, Volkslebenforschung* and *Folkloristik*, its interface with other branches of the humanities and the social sciences, and it states that the present CIAP has failed to satisfy the expectations of the scientific community. The goal envisaged by Ranke is a strong network organisation, capable of initiating and supporting a wide range of loosely formulated cooperation projects and of funding comparative research. The paramount challenge, we are told, was to find ways to finance this expensive program. UNESCO should be approached in a more energetic way, in order to obtain the status of an autonomous institution. If this was unfeasible, the alternative was a closer contact with UISAE, where the three sciences – regional ethnology as practised by CIAP, anthropology and general ethnology – should be on equal footing. Furthermore, a change in the membership principles and the opening up for several categories of members, should secure the finances. The latter proposal would mean that CIAP should be turned into a *Gesellschaft* – a society – instead of a commission with officially elected members.

Through the propaganda of 'the Four' this short and quite unobtrusive document got an almost mythical status, and several folklorists around Europe ordered copies of it.

Except for the rhetorical introduction and the paragraph on the membership principles, there was hardly anything in the *Memorandum* that Erixon could not subscribe to or that he had not himself proposed earlier. Erixon wanted to find a venue for talks with Ranke and the others – in spite of Christiansen's attitude,[81] but it turned out impracticable in the spring and summer of 1962. The first chance to discuss the document was in September in Brussels. There, Erixon had expected that the Voksenåsen committee should have a deliberation before the meeting, as three of them were present, but Ranke declined Erixon's request to gather the group[82] (see above).

In his Report to Christiansen, Erixon claims that Ranke had been appointed also as a member of the new reorganizing committee. This is omitted in the 'minutes' reconstructed by 'the Four'. The question of which document is correct on this point would have been a totally uninteresting detail, had it not been for the role that Ranke actually played in the following period.

The committee work produced a correspondence that is very revealing. The day before their first meeting, in November 1962, Peeters wrote to Ranke to tell him about the forthcoming meeting: 'Your memorandum will provide the basis for our deliberations. I shall not forget to inform you about our first talks before the weekend'.[83] And shortly after: 'It will be a special pleasure for me … to tell you all about our activity concerning CIAP'.[84] Even the confidential information was sent

80 In his private copy, Sigurd Erixon has also underlined '*der letzten Jahrzehnte*' with a big question mark on the margin (SE 8:27).

81 Letter of 3.5.1962 from Erixon to Nettelbladt (SE 8:30).

82 Letter of 12.10.1962 from Erixon to Christiansen (SE 8:31).

83 Letter of 30.10.1962 from Peeters to Ranke (GÖTTINGEN, also PEETERS 8).

84 Letter of 24.11.1962 from Peeters to Ranke (Ibid).

over to Ranke, as for instance the list of persons whom they trusted and wanted to have as officers in the 'new CIAP':

> Enclosed you will find a photocopy of the list of persons whom we have sent documentation on CIAP, as swell as Lecotté's minutes; this is a confidential document, because the names of the candidates are mentioned in it ... We will talk more about it when we meet. I will keep you currently informed about everything that happens concerning our common enterprise [*unserer gemeinschaftlichen Unternehmung*].[85]

The Reporting continued all through the period: 'The 21st of December [1963] we (Lecotté, Pinon and I) will come together for the last CIAP meeting. Immediately afterwards I will inform you about our results'.[86] The only difference between the two letters above is the change from *Sie* to *Du*. There is a strange tone of connivance in the correspondence. As Peeters thought that very much information about the work of 'the Four' should be kept secret, the words *vertraulich* or *confidentiel(le)* are often found in these letters. And Peeters regularly addressed Ranke for advice, almost orders even: 'I would be happy to have an answer from you, which I will forward to the three other members of the CIAP Reform committee, so I can inform them about your advice and your opinion'.[87]

Ranke did not write back so often, but he kept his henchmen informed about important events: 'One of the first days I will send an "open letter" to the president of CIAP. I will duplicate it and send it to a great number of colleagues'.[88] And Peeters Reported happily to the other three when he had received messages from Ranke: '... [Ranke] has not yet found time to write his "open letter" to Erixon [sic], but I think he approves completely of what we are doing'.[89] Or: 'Ranke wants the 4 to come together in November to conclude and to set up the list of names of the adhesions we have received'.[90]

It resorts clearly that 'the Four' regarded Ranke as their commander-in-chief: 'Concerning the letter to Erixon, I [Peeters] do agree with our friend Wildhaber to send a copy to all those who were present in Brussels ... In this way Ranke's position will be strengthened and the open letter will have more effect'.[91]

Occasionally, the commander in Göttingen sent a clear reprimand when he was dissatisfied, as he did to Peeters when Lecotté made a blunder or went too far:

> The reason why I write to you today in all urgency, is a letter I got from Professor Dias in Lisbon, where he informs me that Mr. Lecotté has tried to invite him and his Lisbon colleagues (as well as other European colleagues, I presume) to a meeting in Paris to discuss the organization of CIAP. I consider this very awkward and unfortunate for our

85 Letter of 18.12.1962 from Peeters to Ranke (PEETERS 8).

86 Letter of 9.12.1963 from Peeters to Ranke (MNATP, PEETERS 4). Wildhaber did not take part in this meeting, being in the USA.

87 Letter of 4.3.1964 from Peeters to Ranke (PEETERS 4).

88 Letter from Ranke to Wildhaber of 28.5.1963 (GÖTTINGEN).

89 Letter of 4.6.1963 from Peeters to Lecotté (PEETERS 4). Peeters has written 'Erixon', but it should have been Christiansen – the president of CIAP.

90 Letter of 11.8.1963 from Lecotté to Peeters (PEETERS 4).

91 Letter of 17.5.1963 from Peeters to Pinon (PEETERS 4).

> own efforts [*einen sehr unerfreulichen Querschuss zu unseren eigenen Bemühungen*]. This information has made me very uncertain if I shouldn't just drop the whole CIAP business [*ob ich die ganze CIAP-Angelegenheit nicht in die Ecke werfe*] and devote myself to other, more rewarding tasks. I would appreciate your opinion in this matter.[92]

As will be seen from this and other correspondence, the commando lines followed a fixed pattern: with very few exceptions, Ranke's contact was with Peeters, whereas messages to Pinon and Lecotté went through Peeters. Wildhaber, on the other hand, communicated freely and directly with Ranke.

Without exposing himself, as he would have had to do if he had been a member of the Reform committee, Ranke steered the process and got all information he needed from his willing collaborators. He got Reports, written as well as oral – as they met on quite a few occasions – and copies of important letters, and he gave his advice on policy questions. And he could let 'the Four' attack Erixon and Christiansen as best they could, through open letters and other propaganda activities. As far as can be seen from the epistolary sources, Ranke did not meddle with details nor interfere with the committee's discussions concerning the folklore debate or the name question. Actually, the texts which 'the Four' produced in their Reports concerning the delimitation of the field, the goals of folklore, etc, are so mediocre that Ranke cannot have had anything to do with it. But on a strategic topic like the right – or not – to convene the General Assembly, he told them what they could and could not do.[93]

The campaign of 'the Four' for adhesions was a way of measuring the support. But these lists were also a useful tool for Ranke and 'the Four' when planning whom to invite to the Bonn and Athens meetings.[94] On one occasion, however, Ranke chose to expose himself. The cancellation of the planned Board meeting in Stockholm in May was to his advantage, as it made it possible for him both to expose Erixon and Christiansen to criticism and to come to rescue with a proposal for another venue – where it would be much easier for him to exert control. So early in May Ranke informed Peeters that he found Erixon's explanations for the cancellation 'so groundless' and the deferment of the reform so negative for 'the urgent renewal process' that he would send Christiansen an open letter, in addition to looking for a new venue somewhere in Germany.[95] When Christiansen shortly afterwards was informed of the talks in Stockholm with the CIPSH secretary Jean d'Ormesson (see above), the alarm bells must have rung in Göttingen too.

It was too late now for Ranke to repeat his former proposal for a Board meeting in Göttingen in September 1963. The next upcoming possibility was a German atlas conference in Bonn in April 1964. Ranke stressed that it would be 'a heavy burden to prepare it',[96] but Matthias Zender was willing to cooperate: 'It will be possible for him to invite a number of West European scholars, who would anyway be on our list. I will try to find money to invite the additional people. I am thinking of a group

92 Letter of 6.11.1963 from Ranke to Peeters (PEETERS 4).
93 Letter of 15.6.1963, from Peeters to Ranke (PEETERS 4).
94 See for instance letter of 11.8.1963 from Lecotté to Peeters (PEETERS 4).
95 Letter of 8.5.1963 from Ranke to Peeters (PEETERS 4).
96 Letter of 8.5.1963 from Ranke to Peeters (PEETERS 4).

of 30 to 40 persons'.[97] 'The list' Ranke refers to, contained the names of those who had responded positively to the adherence campaign of 'The Four', and who it might be suitable to invite.

Ranke's open letter to president Christiansen, distributed both in German and French, is dated 2 July 1963.[98] This long letter is an extraordinary text, kept in a polite style and making elegant use of a series of rhetorical effects. It has nothing of the aggressiveness or bluntness of the two prior open letters from 'the Four'. Ranke starts by thanking Christiansen warmly for all his efforts. He then expresses his profound concern for the failure of all these efforts, he praises the work of 'the Four', presents the activities of Christiansen – however well-intentioned they might be – as detrimental, even inimical to the efforts of 'the Four'. He proceeds to draw a list of similar organizations worldwide, claiming their great successes – implicitly asking: why has ours failed? The scenario is that they will take CIAP's place. Ranke then generously offers to serve as a mediator and bridge builder, to find compromises between the two sides, and then comes up with a series of proposals, including an invitation to Bonn.

Finally, having drawn Christiansen down in the mud, he lifts him up again and appeals to his future cooperation and support. In his presentation and handling of the stuff, Ranke balances on a razor's edge. He does never lie, but he plays ignorant where he certainly knew better, and he contributed to the spreading of dubious information where he should know better. In short, Ranke elegantly executed Christiansen publicly, and at the same time appeared himself as CIAP's rescuer. 'An eloquently put letter', was Dorson's comment when it reached the other side of the Atlantic.[99] Erixon's comment, with regard to Christiansen, as cited above, was even more telling: 'I think it's all over with him now'. Other reactions show that the letter had a solid impact on folklorists who had little or no knowledge of the process.

Why attack only Christiansen, when it was Erixon – as Ranke knew very well – who first and foremost had to be neutralised? Erixon enjoyed a deep respect among European ethnologists and folklorists, for his scholarship as well as for his unflagging fight for European ethnology and his many services to CIAP. A public attack on Erixon would have had little credibility and probably a reverse effect. Actually, several persons reacted to the attacks of 'the Four' against Erixon. Erixon was also a feared opponent. In one of their strategic discussions, Pinon actually called Erixon '*la grosse artillerie*' – 'the heavy canon' of the North.[100] So Ranke's strategy was as clear as his letter was eloquent: by discrediting the weakest part of the chain, he would neutralise both.[101] And Erixon, seeing very well the impact of

97 Letter of 14.6.1963 from Ranke to Peeters (PEETERS 4).

98 Letter of 2.7.1963 from Ranke to Christiansen: *Offener Brief an den Herrn Präsidenten der CIAP*. German version (SE 8:27, PEETERS 1), French version (MNATP).

99 Letter of 6.8.1963 from Dorson to Ranke (GÔTTINGEN).

100 Letter of 2.5.1963 from Pinon to Peeters (PEETERS 4).

101 Ranke was careful not to engage in an open conflict with Erixon. When addressing Erixon directly, he was amiable and polite, cfr letter of 21.9.1963 from Erixon to Nettelbladt: '[...] Today I received a very kind letter from Professor Ranke [...] He says that he understands and acknowledges my attitude and my actions, and also that our efforts to help CIAP clearly are well intended and unselfish.' (SE 8:31).

this direct hit, reacted exactly as Ranke must have hoped. As cited above, his remark to Nettelbladt was: 'As things have moved now, it is best to stay neutral'.

A Veteran and his Militant Partisan

One may wonder why Erixon invested so much energy in CIAP. On the practical level, there was a question of securing the projects subventioned by UNESCO. In his letters he shows a deep concern for the reactions from CIPSH to the disorder in CIAP and their threats to withhold or stop the grants. With his earlier experience as a delegate to the UNESCO commissions, Erixon had a better knowledge of the international bureaucracy than many of his colleagues. Also, to carry through an ambitious project like the European atlas would require international coordination of research teams on the national level, that is teams (or national committees) invested with the necessary authority and funding nationally – and not individual researchers.

We meet the same concern about the reduced UNESCO grants in Wildhaber's correspondence, as the printing and the publishing of the bibliography were dependent on these grants. But the bibliography project was to a much higher degree based on a network of individual correspondents. The bibliography project needed the grants, but it could do with a much looser organization than the atlas, hence perhaps the different attitudes to national committees.

To this must be added Erixon's strong conviction that European ethnology, his lifelong vision – as presented and developed in article after article from the 1930s to the 1960s – would be much more difficult to develop as a theoretically-based and comparative discipline, without an international forum which transcended the various sub disciplines. And in Erixon's eyes a firmly controlled organization, with an elected and selected membership, would be a safer alternative for reaching that goal, than a loosely organised society (like CIAP in the late 1940s and early 1950s) where scholars and amateurs alike could become members, and where – in his own words – 'dilettantism' reigned.

Erixon had impatiently pushed Christiansen to arrange the Voksenåsen meeting in 1961, he had asked no less impatiently for the minutes, and when Christiansen was slow in responding, he had written to his colleagues and to his daughter.[102] It was Erixon who spurred Ranke in spring 1962, who wanted to deliberate with Ranke and Roukens in spite of Christiansen's hesitation, who planned the Brussels meeting and who tried to organise meetings in Stockholm in spring 1963. And again it was Erixon who took new initiatives before the Bonn meeting, during the president's silence after the escalation in the summer of 1963.

Ranke wanted Christiansen to summon the April meeting in Bonn, but both the latter and Erixon did not accept it as a Board meeting.[103] Christiansen, who spent the academic year 1963–64 in Leeds and still used crutches, declined the invitation and

---

102 Correspondence with Hilmar Stigum and Elin Christiansen, April 1962, when Christiansen was ill (SE 8:27).

103 Letters of 14.3. and 2.4.1964 from Christiansen to Erixon and Nettelbladt (SE 8:27).

sent a letter, which had been proposed and drafted by Erixon, and which would be read to the participants. The letter welcomed a necessary reform process but warned against dropping the national committees.[104] Erixon decided to participate, partly because Zender's main conference was on atlas questions and he wanted to propose a consolidation of the CIAP atlas commission, and partly because he wanted to discuss the membership paragraph in the proposed bylaw. Erixon hoped to a find way out of the dilemma by proposing a double system, with nationally designated members (to maintain the UNESCO funding) in addition to an open membership system. In cooperation with the Swedish national committee he worked out an amendment proposal, which he sent to Peeters before the Bonn meeting.[105]

In the other camp, the April meeting in Bonn was well prepared. Ranke organised a preparatory meeting in Bonn in early March, with Zender, Roukens and himself present plus 'The Four'. According to the minutes they gave the bylaw a last finish: the name of CIAP was retained – strategically – but "folklore" was used everywhere in the text, while the national committees were kept out.[106] They decided whom to invite – with the list of adhesions at hand. It was by no means an open meeting. Peeters decided not the mention the April Bonn meeting in Report no. 3, or as he explained in a letter to Pinon: 'Because others will ask why they were not invited to that meeting.'[107] At the preparatory meeting they also discussed how to deal with the present Board – whether they should ask it to withdraw or simply regard it as not having a legal basis any longer. Their conclusion was to follow the advice of the CIPSH secretary d'Ormesson, 'to act legally and not make a scandal'. However, the meeting charged Roukens with the task of trying to persuade Christiansen to withdraw before Athens. The only admonition seems to have come from Zender, who warned that the German *Volkskundler* wanted to collaborate with all their colleagues and would not accept them to go too far. The Athens agenda was also discussed; the meeting should start with the dissolution of the old Board, and then a provisional Board should lead the deliberations and the election.

A few days before the Bonn venue (April 27–29, 1964), Lecotté sent a postcard to Peeters. He was alarmed by a memorandum which had just arrived at his desk:

> A memorandum signed Géza de Rohan-Csermak. He's an *attaché* at the *Musée de l'homme*, a friend of G. H. Rivière and very active. Already in Santo Tirso he cultivated his contacts. He talked for a long time with GHR and his thesis closely reflects the views of the latter, who will not hear of anything but Ethnology. I think the author has received our Reports He could have responded and expressed his opinion like others – he prefers to play superstar … we will talk about it in Bonn.[108]

104 Letter of 12.4.1964, signed Reidar Th. Christiansen, President of CIAP (MNATP, Org. Div. – CIAP 1962, 1963, 1964. Several drafts: SE 8:27).

105 Memo dated March 1964 (SE 8:27, Also PEETERS 3).

106 *Comte rendu de la réunion de la Commission de Réforme de la CIAP tenue à Bonn les 7 et 8 mars 1964* (PEETERS 3).

107 Letter of 13.1.1964 from Peeters to Pinon (PEETERS 4).

108 Message of 17.4.1964 from Lecotté to Peeters (PEETERS 4).

Géza de Rohan-Csermak's entry on the scene came to mean a good deal to Erixon, especially from 1965 onwards, when they started a close cooperation. The unreserved support of the young and rather militant de Rohan-Csermak (1926–76) seems to have spurred the old veteran to a continued fight. Among other things, de Rohan-Csermak should become the co-founder and first editor of Erixon's journal *Ethnologia Europaea*. Several persons were positively surprised when de Rohan-Csermak's memorandum unexpectedly appeared. Or as Brantanic wrote to Erixon: 'It was a pleasure, for once, to read something intelligent. Don't you think that here we see the emergence of a young *Secrétaire Général*?' [109] On the other hand, Pinon was deeply offended by this '*super-réformateur*' who showed so little respect for the Reform committee. He immediately made a detailed analysis of the document.[110] He had to admit that it had certain qualities, but he dismissed its ethnological and anthropological profile.

The memorandum, *Pour une association d'ethnologie européenne*, is a well structured and coherent document of 46 pages, dated 'Easter 1964'.[111] The first half of the document gives a historical overview of European ethnological cooperation since the end of the nineteenth century (and this overview impressed even Pinon), discusses the different traditions, as well as the relation between European ethnology in Europe and the study of European heritage and traditions worldwide – thus bridging the European and the global and situating European ethnology as a global discipline. Its aim is a synthesis of European and non-European ethnology and a closer integration of the three main study objects of ethnology: spiritual culture, material culture and social culture.

The second half discusses practical issues of a reform: the name, the relation to UNESCO and other organisms, a future journal and the bibliography, as well as questions of financing. As for the bylaw, de Rohan-Csermak gives – somewhat reluctantly – his acceptance to the one proposed by 'The Four', on condition that *ethnology* replaced *folklore*; since the term for a field of study could not design a scientific discipline. He keeps a neutral tone, but when discussing the problem of terminology and folklore, he cannot refrain from a jump on 'the Four', for 'imposing their personal opinion ... on a whole science' and to have engaged in 'an aggressive and impolite diatribe against a respected professor of ethnology' (i.e. Steinitz, cf. pp. 44–5).

At the April meeting in Bonn, with about forty scholars present and Roukens presiding, this memorandum – by all standards the most coherent and scholarly document produced in the reform process – was quickly dismissed with two remarks: it was not an official document, and it coincided on several points with the proposals of 'the Four'.[112] There is no trace in the minutes of a debate on Erixon's

109 Letter of 18.4.1964 from Bratanić to Erixon (SE 8:27).

110 *Examen du Memorandum présenté par Géza de Rohan-Csermak en vue d'une réorganisation de la CIAP* (PEETERS 3).

111 (MNATP, Org. Div. CIAP – *Réunion de travail*, Bonn 26–27 avril 1964; Nordiska/SE).

112 *Compte rendu de la Réunion tenue à Bonn les 26 et 27 avril 1964 en vue d'examiner les dernières propositions de la Commission de Réforme de la CIAP* (SE 8:27. Also MNATP).

compromise proposal for a membership system with both individual members and national delegates. The name CIAP should be kept. The most noteworthy change is that *folklore* was replaced by *vie et traditions populaires* – 'folk life and folk traditions' – in the bylaw. This was explained in the minutes as 'a compromise to end all the quarrels on the name of the discipline'. And the Reform commission – 'The Four'– was prolonged, to prepare the Athens meeting.

'The Four' proposed a 'CIAP *universelle*', a worldwide organisation, based on continental sections. However, a majority in Bonn found this to be too difficult; among other problems, there existed already the South American CIFP (see above). It was decided that only a European section should be started in Athens, but the bylaw should be formulated in a way that would permit a later establishment of continental sections. As UNESCO would not support a purely European organization, it was now clear to all that there would be no more funding from CIPSH.

The relations to Eastern Europe came more into focus in this late phase of the reform process. In the 1950s, contacts seem to have been rather few across the Iron Curtain, and after the war CIAP had for most purposes been an organization for Western Europe. Only one Eastern country had been represented on the CIAP Board, when Prof. Xavier Pywocki (Warsaw) was elected in 1955. But with the decline of CIAP in the following years he hardly took part in any activities. Contacts were better with the much freer Yugoslavia; Milovan Gavazzi (Zagreb) was on the Board since 1954, and since the Amsterdam congress in 1955 Bratanić joined the cartography commission. During this congress, however, an East-West-incident created a cold front to the DDR. Dr. Oskar Loorits, Estonian refugee to Sweden, accused '*der rote Professor*' – Wolfgang Steinitz (East Berlin) – for espionage, provocations and threats of execution, which had forced Loorits to flee his country. Loorits' accusations and his plea to CIAP to come up with a policy against political oppression of scholars and a support to exiled folklorists created much turmoil during the congress and rather strained relations for a period.[113]

The same Steinitz attacked 'the Four' in 1963 for their narrow concept of folklore; he stressed the role of material culture studies in *Volkskunde* and the need for keeping close contacts with IUAES (see above). After the Bonn meeting, however, Steinitz came to function as a bridge builder between 'the Four' and his Eastern European colleagues. In the summer of 1964, Steinitz Reported to Peeters and Pinon that he had registered reluctance to the CIAP reform among Soviet, Bulgarian and other Eastern European colleagues. One of their concerns was the reformers' demarcation against the anthropological sciences. At the ICAES (IUAES) congress in Moscow the same summer, where folklore and European ethnology were allotted several sessions, CIAP and its reform was discussed. The Eastern Europeans did not hail the idea of a rival organization to IUAES. Steinitz explained that he had done his best to convince his colleagues, but he advised 'the Four' to establish better contacts with Eastern Europe, and even to invite a person from a socialist country to join the Reform committee.[114] 'The Four' saw the point of an East European campaign

113 See correspondence (Meertens 35:1131; Arnhem 558). See also Steinitz' response in *Deutsche Jahrbuch für Volkskunde* II (1956)

114 Letter of 18.7.1964 from Steinitz to Pinon and Peeters (PEETERS 3).

immediately, and established contact with both Gyula Ortutay (Budapest) and Mihai Pop (Bucharest).

The weak contacts with Eastern Europe during the Cold War had several side effects. At the abovementioned Budapest congress in October 1963, Matthias Zender had learned that Russian and other Slavonic ethnographers were planning an atlas of vernacular architecture and agricultural tools, to cover Eastern Europe and the contingent regions, i.e. the belt stretching from Finland over West Germany and Yugoslavia to Greece, but excluding Scandinavia and Western and Southern Europe. Zender had recently taken a seat in CIAP's cartography commission, after the decease of Richard Weiss (Switzerland). Worried about this new frontier, which crossed the area of the ongoing West European atlas work, but also seeing a possibility of a Pan-European atlas covering the vast area from the Atlantic to the Ural, he informed Erixon and the rest of the cartography commission.[115] This is probably the reason why Zender, even after his commitment for 'the Four' in Bonn in 1964, did not want to take on an office in the 'new CIAP' – or as Lecotté explained in a letter: 'Zender has declined, in order not to complicate his relations with the East'.[116]

After the Bonn event both camps invoked the right to convene the meeting in Athens. Christiansen planned an agenda, and so did 'The Four'. And both sent out summons for the same venue; the president and the secretary to the national committees and other official delegates, and 'the Four' to the scholars on their adherence list. When Nettelbladt sent out invoices for the fees (for the years 1959 to 1964), in order to secure as many legal votes as possible according to the (old) bylaw, 'the Four' regarded it as a 'manoeuvre'. And when Christiansen applied to CIPSH for support for travel costs for Board members and other participants, 'The Four' intervened in order to stop the grant. CIPSH secretary Jean d'Ormesson, wanting a reform but insisting on legal procedures, must have felt the situation rather delicate.[117]

On 27 July, the secretary Nettelbladt knocked on Rivière's door in Paris. The topic of their conversation was whether Géza de Rohan-Csermak could become general secretary of CIAP – an idea that Bratanić had launched already in April. Csermak was willing, but Rivière's acceptance was necessary, as the latter was Csermak's supervisor for his thesis.[118] Rivière gave his acceptance, and Christiansen immediately appointed Csermak interim general secretary, to replace Nettelbladt at the assembly in Athens. In late August Csermak went to Stockholm, where he had talks with Erixon and Nettelbladt. Erixon, who had first been hesitant to this solution, changed his mind after Csermak's two days' visit to Stockholm. 'We found him wise, very balanced and diplomatic', Erixon wrote to Brantanic.[119] And Dias, who had met him at the Santo Tirso congress, was no less enthusiastic: 'He is young, intelligent, and ambitious; he speaks foreign languages and – what is especially important – he

115 Letter of 24.10.1963 from Zender to Erixon (PEETERS 4).

116 Letter of 25.5.1964 from Lecotté to Peeters (PEETERS 6).

117 See correspondence (PEETERS 3,5,7, SE 8:27). In total 40-50 letters from August/ September 1964.

118 Memo of 27.7.1964 (MNATP, Org. Div. CIAP–SIEF 1964-65-66-67).

119 Letter of 26.8.1964 from Erixon to Brantanic (SE 8:27).

has a broad and scientific outlook on ethnology and he does not share the narrow and limited conception of the folklorists'.[120]

The Athens meeting was imminent, and Csermak did not waste his time. The proposition of an alternative bylaw, presented in the name of Christiansen, was his work. The main amendments are the suppression of continental committees and the reintroduction of national committees, supplemented by individual membership – i.e. Erixon's proposal. And he obtained from the CIPSH secretary d'Ormesson economic support for the travel of the Board members. The latter could not resist adding that he had now 'received from several sides initiatives, the strict legality of which might be questioned'.[121] For various reasons Erixon, Rivière and Dias – three central actors holding deviant views on some of the main issues – could be present in Athens. Correspondence in the weeks preceding the assembly gives a picture of what was at stake.

In a letter to his deputy, Marie-Louise Tenèze (MNATP), Rivière gives directives as to how she should vote.[122] Concerning the terminology, she should accept the compromise *folk life and folk traditions*, which replaced *folklore*. She should repeat Rivière's proposal from Bonn that CIAP's close relation to UIAES should be specifically mentioned in the bylaw. And she was ordered to propose the reintroduction of national committees, with a motivation that coincides with the view of Erixon – beyond that of the purely economic (UNESCO) considerations: 'I still sincerely believe that the only way for an organization, with high ambitions to develop with the desired efficiency and equity, is to structure itself on the basis of national committees; otherwise the risk is to end up with congregations and sects.'

Rivière's fourth admonition concerns the Reform committee's proposal of continental committees, that is the proposed umbrella structure that should make CIAP (at a later stage) a worldwide organization, and thus satisfy UNESCO's requirement of being global. To him, as well as to Erixon, Dias and de Rohan-Csermak, the universal aspect of CIAP should derive from scientific considerations and grow out of its object of study, and not be linked primarily to an administrative structure. The idea of CIAP trying to create something parallel or rival to IUAES was to him both unrealistic and foolhardy.

Erixon, who had no special interest in joining the folk narrative conference in Athens, found the travel from Stockholm to Athens too long and expensive. Like Rivière, he gave priority to the IUAES (ICAES) congress in Moscow in August – where atlas questions were among the topics to be discussed. In Moscow close links were knit between CIAP's cartography commission, IUAES and Russian researchers, with a view to cover all of Europe. For Erixon, as for Rivière, it seems that IUAES had become just as interesting as CIAP. Furthermore, Erixon was by no means sorry for the strong interest among Russian cartographers for material culture issues. In letters to other cartography specialists – Vilkuna, Dias and Bratanić – he reveals that he was in doubt whether the cartography commission should be

120 Letter of 2.9.1964 from Dias to Erixon (SE 8:27).

121 Letter of 26.8.1964 from d'Ormesson to Nettelbladt (SE 8:27).

122 *Directives remises à Mme M. L. Tenèze, Chef du département ...*, of 28.8.1964. (MNATP, Org. Div. CIAP–SIEF 1964-65-66-67).

considered a CIAP commission any longer – as no support for their meetings had been given through CIPSH since 1954.[123]

In the weeks before the assembly in Athens, Erixon exchanged views with several actors. He was preoccupied with formal aspects of voting, as most of the Board members would be absent. Bratanić and Erixon carefully analyzed the situation, concluding that it would be difficult, perhaps impossible, to carry through a General Assembly in a fully legal way; there would be too few present with a vote, if the bylaw in force should be followed, to reach a valid decision. Some sort of compromise had to be sought.[124]

Erixon had not quite given up the hope of getting CIAP on rails again, so he asked – in agreement with Christiansen – Dias to stand for the presidency. They could accept Ranke as vice president; they would even – 'in an emergency'[125] – accept him as president, if Dias was willing to be vice president. Erixon may have had his suspicions, but he could not know – as little as anyone else – about the very close bonds between Ranke and 'The Four'. Their candidate for the position of general secretary was Géza de Rohan-Csermak. Dias answered that he agreed fully on the choice of secretary, but feared that 'the consolidated folklorists will try to fill all important positions in CIAP'.[126] As for his own candidacy, he replied that the desire to be president for 'a folkloristic CIAP' was indeed small, and that he would never accept the vice-presidency under a folklorist president ... this would force me to unproductive compromises ... and deprive me of my liberty to perhaps join an effort to establish another European ethnological organization, perhaps with financial help from some foundation or other.'

To conclude, what was at stake, and more or less explicitly discussed, was:

- the membership basis – national committees/delegates or individuals;
- the geographical scope – Western Europe, Europe or the world;
- a universality founded on formal and administrative criteria (a federation of continental sections) or on scientific criteria (methodology and subject matter);
- the division of labour and the organizational relations to the anthropological sciences, and to other international organizations;
- the definition of the discipline – folklore in a restricted sense, or material and social culture as well;
- the hegemonic strife between folklore and ethnology.

This is not to say that there was an absolutely clear-cut division between two camps. To some extent, there were crossing and intersecting interests. Not everyone on the adhesion list would subscribe to all that 'the Four' proposed.

123 Letters of 22.8, 26.8. and 27.8.1964 from Erixon to K. Vilkuna, Bratanić and Dias (SE 8:27).

124 See especially letter of 1.9.1964 from Bratanić to Erixon (SE 8:27).

125 Letters of 26.8. and 27.8.1964 from Erixon to Bratanić and Dias (SE 8:27).

126 Letter of 2.9.1964 from Dias to Erixon (Nordiska/SE 8:27).

The Winner takes it all: Athens 1964

The CIAP Board meeting and the General Assembly were strategically placed on 7 and 8 September, in between a congress of the International Society for Folk Narrative Research and a workshop on a folklorist book series. With no safe legal procedures to follow, there was a strong tension in the air. The most nervous person was probably Peeters, to judge from his personal notes, with short PMs in the margins: 'The 4 cannot make a fool of themselves', 'If the 4 aren't unanimous, all has to be redone', 'I have contacted lawyers', etc. Some were probably discrete messages to Pinon or Lecotté during the deliberations.[127]

The sources to what happened in Athens are of two kinds. First there are the official minutes, later printed in *S.I.E.F–Information* no. 1. These give short notices of the voting and the decisions taken, but do not contain any information about the debates, the issues at stake, and the train of events. The same applies to some Reports published in scientific journals, like Beitl 1964. The other type of source is the first-hand witness Reports, from central actors present at both the closed and open meetings – in this case long and very detailed Reports from Bratanić and de Rohan-Csermak. The official minutes are the winners' version, certainly correct as for the formal decisions taken, but leaving out all that is not strictly necessary to broadcast. Their lack of objectivity resides in all that is left out. The two first-hand Reports add and thus rectify, but as the losers' version they are no less arbitrary and subjective than the official sources. When I find it necessary to remind of this obvious fact, it is because the eyewitness Reports from Athens unveil a series of events, adversities and tribulations, which were little visible to those who did not participate in the closed meetings, and hidden to the scholarly community at large.

The official minutes from the Board meeting are only a few lines, but the abovementioned Reports narrate in detail what happened. Actually, it was not an ordinary Board meeting, as only three Board members were present (Christiansen, Roukens, Thompson), two per delegation (Bratanić for Erixon and Gavazzi, and Znamierowska-Prüfferowa for Piwocki), and de Rohan-Csermak as interim secretary. In addition, Ranke and 'The Four' participated in all sessions. The main issue was the negotiation of procedures for the General Assembly, including several attempts from 'The Four' to disallow the authority of de Rohan-Csermak. The latter reports on Peeters' feats of rage and very blunt attacks during the closed sessions, episodes that are confirmed by other letters, and of Christiansen's threatening to leave the meeting should Csermak be expelled. In his private papers, Peeters have noted sentences like: '150 pers[onal] and instit[utional] adhesions, Bonn approval from 40 persons representing 17 countries – and now when all is ready R[ohan] C[serma]k turns up', 'The General Assembly should decide whether it's the Committee of 4 who should assist Christiansen, or is it R. Ck.', etc.[128]

At the General Assembly, 'The Four' took the direction; they seated themselves at the presidential table together with Roukens. Csermak had to step down and take

127 Peeters' handwritten notes for/from the meetings (PEETERS 5).

128 Letter of 26.9.1964 from de Rohan-Csermak to Nettelbladt. See also letter of 14.10.1964 from Nettelbladt to Peeters, and Peeters' drafted response (PEETERS 7).

a seat with the assembly, while Christiansen chose the role of an observer. The new bylaw was voted first, and then followed the election of the president and the members of the Administrative Council, and finally a voting over the name of the organization.

During the bylaw debate, a proposal to combine national delegates with personal members was voted down, whereas a system with continental commissions was reintroduced. Contrary to the bylaw, Roukens and 'the Four' demanded the election of the president first – and Peeters was elected. Among the fifteen members elected to the Administrative Council were Wildhaber, Roukens, Zender and Ranke, whereas Bratanić did not get sufficient votes. The Administrative Council then constituted itself, with the result that Mihai Pop, Carl-Herman Tillhagen (Stockholm) and Richard Dorson became vice-presidents, Pinon general secretary and Lecotté treasurer. This meant that Peeters, Pinon and Lecotté now formed the core of the executive Board; after two years of work they had reached their goal of filling these positions.[129] The leading team of six persons were all folklorists in a restricted meaning. Critics like Erixon and Rivière were appointed honorary members.

The new Council decided to break the Bonn compromise to keep the name of CIAP. They summoned the Assembly again and launched a new name for a vote. After a close race (17 against 15) *la Société Internationale d'Ethnologie et de Folklore* (SIEF) replaced *la Commission d'Art et Tradition Populaires*. Ranke had now got his 'society' in the name, and Pinon and Lecotté their 'folklore'. This was a blow to all those who wanted a closer collaboration with the anthropological sciences, IUAES and CIPSH. The old name had given a hope of continued contact, but a new name as well as a new organization signalled a full break and would not be accepted by CIPSH. It was also a strike against those who were of the opinion that folklore was an integral part of European ethnology, as were Erixon, Dias, Rivière, Bratanić, Csermak and several others. There is more to a name than just the name, in other words.

During the General Assembly a rather chaotic situation had arisen concerning who had voting rights, as no members were yet registered. In the end, those who were present were allowed to vote, while delegations from absent members were not accepted. There were some irregularities in relation to both the old and the new bylaw, i.e. the refusal to use the delegations and the sequence of the elections. Csermak drafted a letter of complaint to CIPSH, but the older and wiser Bratanić dissuaded him from sending it.[130]

After the Athens meetings, Csermak wrote a long Report (22 pages),[131] and Bratanić a somewhat shorter one in addition to a comment to Csermak's Report (10–12 pages), where he confirms the latter's version of the events.[132] These

129 'The Four' had left their earlier idea of Dorson as president. With the new plan for continental commissions, Dorson was intended for the North American commission. On several later occasions, Lecotté had announced Peeters' candidacy. Letter of 26.9.1964 from de Rohan-Csermak to Nettelbladt (PEETERS 7).

130 Letter of 23.12.1964 from Bratanić to de Rohan-Csermak (DIAS Box 4).

131 Letter of 26.9.1964 from de Rohan-Csermak to Nettelbladt (PEETERS 7).

132 Letter of 23.12.1964 from Bratanić to de Rohan-Csermak (DIAS Box 4).

documents are so detailed that it is very difficult to render them full justice in this article. Csermak describes in detail attacks and personal insults he suffered from Peeters, and what he calls 'the putsch'. The bitterness of Csermak's text may be rendered through a few sentences from the conclusion of his long letter:

> We have certainly been 'dupes'. But under no circumstances would we have accepted to use the same arms. To conduct electoral campaigns, to spread rumours, to go from door to door to win votes – that is the job of village politicians, but it is unworthy for us who have science as our vocation. [...] We were ready to discuss any theoretical or methodological aspect of our research. But we were not good at defending us against attacks that came more often from behind than in front, by these 'Sturmabteilungsmänner' (p. 21).

Bratanić's shorter Report[133] is more interesting in the sense that his presentation differentiates between a descriptive, an analytical, and an assessing part. It contains an election analysis which indicates that the tactical steps – cf. the membership question and the voting sequence – influenced the result of the elections. It also becomes clear from his Report that Wildhaber and Ranke were the dominating voices at the General Assembly, and that it was Ranke who gave 'the Four' the necessary authority. As Bratanić was less personally involved than Csermak, not being the target of personal attacks, it is worthwhile to listen to a longer quotation from his § 3, entitled 'My personal assessment':

> My personal assessment of all these events is very negative. I cannot hide that the way the Assembly was conducted and the behaviour of the leading 'Reformers' disappointed me very much. By the leading 'Reformers' I mean not only the 4 members of the Reform Commission, but also professor Ranke and perhaps even some others. Their behaviour towards the old Board and towards all who did not unconditionally and in all questions side with the 'Reformers' cannot be described in more positive terms than unfair. The (apparent and supposed) errors and weaknesses of the old Board were persistently regarded as unforgivable, whereas the much more important errors, awkwardness, even unacceptable behaviour of the Reform Commission have evidently been considered acceptable and normal ... By spending very much time and other resources the 'Reformers' have directed all their energy – which is not little – towards a small and narrow target: to take over completely and unconditionally the organization of CIAP ... Already the displacement of the venue to Athens, to a congress treating a very narrow topic, and only one month after the big [IUAES] congress in Moscow, where a much greater number of interested colleagues – ethnologists as well as folklorists – participated (4 out the 8 that gave me procuration), made one suspect that it was the intention all the time to reduce the CIAP Assembly to a small and specialised circle. This impression was strengthened by the list of preliminary new members of CIAP [the adherence list] (for a large part folklorists in its narrow sense) ... and the very bizarre methods of recruiting these members (p. 5–6) ... In this way a 'voting machine' was organised in Athens ... With such methods it was not difficult to defeat those who were unprepared, those who trusted the force of arguments and had more important things to do than to prepare themselves for such a 'fight' and 'power usurpation'. The most depressing, and the worst of all, is that this petty and narrow-minded fighting spirit is without any real purpose (the introduction of the term 'ethnology' in the name of the organization – a term that they have earlier attacked aggressively – *after*

133 Letter of 25.19.1964 from Bratanić a series of colleagues (SE 8:27).

their 'usurpation of power', is a good example of the lack of principles in this 'fight') ... Anyway, the result is that not only are all the members of the Reform Commission voted into the Administrative Council, but the new Board is composed *exclusively* of them (Peeters, Pinon, Lecotté). These three men have now according to the new bylaw practically absolute power [*die Alleinherrschaft*] over the new organization for 3 years. Their position up to now, their behaviour and their declarations give no reason to presume that they will be capable – not even partly – of standing up to what must unconditionally be expected, concerning scientific as well as personal qualities, from the leaders of an international organization like ours.

There was another *Volkskundler* who was concerned with *realpolitik* and who had his misgivings – Matthias Zender. Zender had supported Ranke and 'The Four', but he had also warned against going too far (the Bonn meeting in March), and he was concerned with the cooperation with Eastern Europe. Only a few days after the event in Athens he wrote to Peeters:

Even if our colleague Bratanić did not get my support in Athens, mainly because of his collaboration with Csermak, I strongly regret that he was not elected to the Administrative Council. Above all Bratanić has rendered meritorious services to us in our many hard disputes on the implementation of the European atlas. All the West European atlas projects have much to thank him for. So I think it would be right, in spite of his partial opposition, to offer him a seat in the Council. In this way we may avoid that he makes difficulties for us and for himself, in his state of anger.[134]

Zender offered to step down from the Council and let Bratanić have his seat, if Peeters could find a way of doing it. Peeters' responded by proposing to make Bratanić the leader of the cartography commission; thus he could formally attend the Council's meetings.[135] However, Peeters should soon learn that this was reckoning without one's host.

If Csermak was furious, Bratanić was disappointed and Zender had his qualms, there was at least one who felt a great relief – Winand Roukens. Roukens had worked very closely with Ranke and 'The Four', all the way from Voksenåsen, via Brussels and the two Bonn meetings to Athens, where he had presided the plenary sessions efficiently – so efficiently that Rivière sent a disapproving letter to Peeters because Roukens had not even put to a vote his and Marie-Louise Ténèze's formal bylaw proposal on the IUAES connection.[136] But Roukens himself felt so relieved on his way home from Athens that he wrote a Goethe-inspired poem – *Europa-Gedanken* or 'Meditations on Europe' – dedicated to Kurt Ranke.[137] There is little doubt as to whom Roukens considered the architect behind SIEF.

---

134 Letter of 17.9.1964 from Zender to Peeters (PEETERS 8).

135 Letter of 12.10.1964 from Peeters to Zender (PEETERS 8).

136 In the same letter Rivière declined to accept his seat in the new SIEF Council. Cf. letter of 24.9.1964 from Rivière to Peeters (PEETERS 8).

137 *Europa-Gedanken (nach Athènes ...). In Freundschaft für Prof. Dr. Kurt Ranke*, signed Winand Roukens, *Athen-Delegierter der Universität Nijmegen*, Strasbourg 17.9.'64 (PEETERS 8). It is a rather pompous poem, rephrasing Goethe, on building bridges between peoples and constructing the European house – in the name of humanism.

Aftermath

In spite of his disappointment with the result in Athens and his harsh words in the letter (cited above), Bratanić accepted the defeat. When Dias and even more eagerly Csermak soon after mentioned the possibility of founding a new organization, Bratanić warned – not against a new organization, but against a *rival* organization. In December 1964 he wrote to Csermak:

> All this need not hinder us from imagining a new and *better* organization, and when time is ripe, also of realizing it, 'without breaking in a brutal manner with SIEF' (and as we talked about in Athens, this new organization could be of another type than CIAP/SIEF; that is, not necessarily a competing enterprise). Our objective should rather be the *creation* of something positive, and not to fight against something.[138]

Erixon did not want a re-enactment. In October 1964, he explained to Nettelbladt that he 'accepted the ruin of CIAP', but he wanted to continue his work in the commissions.[139] On one point he even admitted a certain relief; 'I have told this to several of my foreign contacts and I do not any longer need – for kindness' sake – to defend the neglects of Christiansen.' This was hardly an exaggeration, as he had often written in his letter drafts that he was impatient with Christiansen's slow tempo – remarks which either he or his translator used to drop in the final version of the letters. However, when he added that he had only been a councillor, and 'partly in secret', it is not the full truth; he gave advice currently, but not infrequently had Christiansen's letters been both conceived and drafted by Erixon.

Christiansen, who felt that a burden had been taken from his shoulders, willingly signed a declaration of succession, stating that SIEF was the continuation of CIAP. Peeters wanted this declaration in order to convince CIPSH that SIEF should inherit all rights after CIAP. Because he feared that Erixon or others would try to revive CIAP, Peeters even published the declaration in the first issue of *S.I.E.F.-Information.*

Nothing indicates that Erixon planned to start a new organization, even if Peeters repeatedly suspected this in his letters. But in autumn 1964, Erixon began exploring the possibilities of launching a new journal for European ethnology, and in spring 1965 he convened Dias, Csermak and Bratanić to a planning meeting. He had no objection to continuing the work on the dictionaries under the auspices of SIEF, but he was more reluctant about the cartography commission, which had recently strengthened its bonds to IUAES. The result of the 1966 meeting in the commission was that it came to regard itself as independent of SIEF – as *die Ständige Internationale Atlaskommission.* Bratanić even declined a request to invite Peeters to its conference in Zagreb in 1966, a conference which attracted participants 'from almost all European countries'.[140]

This was not the only blow to SIEF. Except for the production of the first volume of *S.I.E.F.-Informations* (vol. 1/1964), the activities started out on a low pitch. CIPSH

138 Letter of 23.12.1964 from Bratanić to de Rohan-Csermak (DIAS Box 4).

139 Letter of 17.10.1964 from Erixon to Nettelbladt (SE 8:27).

140 Letter of 17.11.1965 from Bratanić to Mauritz de Meyer (PEETERS 11). This refusal was badly received by Peeters and led to some correspondence.

would not give any grants to SIEF for 1965–66, and the dialogue with UNESCO turned out difficult. The treasurer Lecotté complained in April 1966 of 'the lack of harmony between the received adherence forms and the incoming payments';[141] the collecting of fees from the members should once more turn out a demanding task, and not less so with a high number of invoices and a low fee. SIEF faced considerable economic problems the first years; its officers still had to pay privately their travels, and for every meeting there had to be sought external financing.

The envisaged series of congresses and the planned scientific journal were quickly put on ice. There had been a general agreement that a newsletter would be of paramount importance for creating a lasting interest for SIEF. But even the publishing of the newsletter – the *S.I.E.F.-Informations* – turned out an almost insurmountable task. In the two first years there was no money to have it printed, nor an editor could be found; Herman Bausinger, Rudolph Schenda, Lauri Honko and others were asked, but they all declined. The newsletter had been intended as a quarterly publication, but it took one and a half year before no. 2 appeared. Only four issues were published between 1964 and 1970, when the last one in this series appeared. The person who finally accepted to do the editing work was Wilhelm Nicolaisen (Edinburgh), and for economic reasons the printing was done in Budapest. In the late 1960s, the problem was not primarily the economy; according to Nicolaisen's correspondence with the Board it was impossible to get in enough contributions to publish even the newsletter.

The planned first Council meeting in Antwerp had to be postponed and organised in conjunction with the congress of the *Deutsche Gesellschaft für Volkskunde* in Marburg in April 1965, with a follow-up Council in Antwerp in September. But very little happened and the climate worsened. The situation after the first year can best be described through Wildhaber's outburst in December 1965, when he complained to Peeters over the inactivity of the general secretary and the first vice president:[142]

> I must write and tell you that I am very worried about SIEF ... Erixon and Bratanić take initiatives and they are working; we are making plans, and nothing happens! By the end of September Mihai Pop was charged with the task of writing to all the commissions and ask if they would adhere or not. *Nothing* has happened yet! How can we 'Report on the activities of the commissions' in Prague [item on the agenda of the forthcoming Administravie Council, planned to take place in Sept. 1966], when these *de facto* and *de iure* do not even exist yet! The 'others' will laugh and be merry, and they will say that they are working, but how about us? On my recent journey, some of my acquaintances asked me whether they really belonged to a commission or not; I could not give them an answer.

Pinon, who was also criticised repeatedly by the rest of the Board, explained the situation in this way:

> I must repeat what I have already said: for a person who does not dispose of a staff or a secretariat, it is a too heavy a task to keep going the general secretariat of SIEF. It is either this secretariat and intellectual sterilization, or intellectual work and a secretariat that

141 Letter of 14.4.1965 from Lecotté to Peeters (PEETERS 6).
142 Letter of 2.12.1965 from Wildhaber to Peeters (PEETERS 8).

languishes. I haven't been able to do better for you. It's up to the Administrative Council to reflect upon it and find a better solution.[143]

To Peeters, the difference between being in opposition and in position could hardly have been greater. After only one year, in the autumn of 1965, he signalled that we wished to resign. Pinon was periodically ill – or silent. In spring 1967, Lecotté fell ill and Pinon left for a year in Bloomington. The membership was rising quickly, and with as many as 400 members for a short period (a third of them from USA), there was more than enough to do. The problem was that also the number of complaints from the members was rising – for not receiving any information nor any newsletter. No one wanted to take over the position as general secretary, even temporarily in Pinon's absence, so Peeters had to take over the secretarial tasks himself for a period. From 1968 Mihai Pop took on the secretariat, and Pinon was forced to resign as general secretary. Peeters announced several times his wish to step down from the presidency, but he was forced to remain in the seat until 1971 – not only for lack of candidates and because they did not manage to organise a General Assembly, but also because Ranke would not accept changes in midst of the first period – 'otherwise we will lose the last rest of prestige we still have.'[144]

In the camp of 'the others', to borrow Wildhaber's expression, the climate was quite different. With the assistance of de Rohan-Csermak, Erixon organised an important international conference in 1965, at Hässelby near Stockholm, where theoretical and methodological aspects of European ethnology were discussed. The conference also decided an offensive programme, with yearly conferences, a scientific journal and detailed plans for a series of ethnological handbooks. The second conference was held in 1966 at Julitta, also near Stockholm, and the journal appeared in spring 1967: *Ethnologia Europaea. A World Review of European Ethnology*, vol. I contained an impressive overview of the research and teaching of European ethnology worldwide (a task which also one of the new SIEF commissions had taken on, with no results so far). Except for a few scorning remarks from its editor, Csermak, SIEF is not mentioned at all in this overview. In his editorial article 'European ethnology in our time', Erixon reduces CIAP/SIEF to one single phrase – 'Much has been said about the activity of the CIAP after the last war' – and then moves on to a presentation of IUAES (p. 5).

At Hässelby the founding of a formal organization was discussed, but it was decided to follow Bratanić' advice to remain a loosely organised network (Csermak 1967).

In addition to these conspicuous successes, the 1966 Zagreb cartography conference organised by Bratanić had come up with detailed plans for a new organization of the European atlas work. (Bratanić 1966, 1967) The conference had ended on a very positive pitch and a series of yearly congresses was announced. That the plans were extremely ambitious and too optimistic as to what cartography could contribute to ethnology, is another story; the independence and apparent success of

143 Letter of 14.1.1967 from Pinon to Peeters (PEETERS 7).

144 Letter of 22.12.1965 from Ranke to Peeters (PEETERS 8).

the *Ständige Internationale Atlaskommission* (SIA) was hard to swallow for SIEF, which had planned its own commission from the start.

So Wildhaber was right: 'the others' worked, whereas nothing happened in SIEF. An additional headache, as discussed between 'The Four', was that central SIEF members (Dias, Zender, Meertens ...) participated also in the Erixon-Bratanić-Csermak network. And Pinon Reported from USA that *Ethnologia Europaea* had aroused much interest and that American folklorists expected something more than silence in return for their yearly fees to SIEF.

The minutes from the two SIEF Council meetings in 1966 (Göttingen and Prague) are held in a remarkably humble and reconciliatory tone. At an extraordinary crisis meeting in Göttingen in March the Council discussed how to bridge the gap between folklore and ethnology and 'bring to an end the outdated distinction between spiritual and material culture'. The Council decided to approach Erixon by means of 'a manifest on the new orientation developed through the discussion', to be worked out by Ranke, Dias and Nicolaisen. And letters should be sent to Erixon and Bratanić. None of the decisions seem to have been carried through. At the Council's meeting in Prague in September the discussion continued, and there was also a debate on SIEF's focus on 'European culture' and the term 'European' – a term which Ranke now wanted to see in the name of the organization.

In 1968, SIEF was accepted as a member of IUAES – a 'subcommission' status that 'The Four' had strongly opposed – and thus affiliated with CIPSH/UNESCO from 1970. SIEF was not accepted as a CIPSH member in its own right, which had been a claim after Athens. The condition for being recognised by IUAES had been a clear division of labour – which meant a renunciation of SIEF's worldwide ambitions and the plan for continental committees. The agreement upon the European profile at the 1966 Prague meeting must be understood in this context.

There is a discrepancy between the official documents (minutes) and the correspondence. In the latter current problems are discussed openly and complaints and criticism of the low level of activity are rendered; the letters even give glimpses of internal quarrels. Ranke was no less worried that the others. It was him who had found money for the extraordinary Council meeting in Göttingen in 1966 (above) to discuss the critical situation in SIEF. But the state of affairs remained largely the same, and early in 1967 Ranke wrote to Peeters:[145]

> What shall we now do for SIEF in the future? I think that we, for our part, have done everything that was possible, in order to give this young organization life, vitality and the force of credibility. But for my own part I must admit – and you will agree with me – that except for council meetings and designation of commissions – from which we have heard nothing – nothing has happened yet. It's enough to drive you to despair [*Es ist zum Verzweifeln.*] Perhaps we should meet once more, a few of us, say in Bonn, only three or four persons, and talk over these things, before the Council is summoned.

The clearest symptom, in addition to the conspicuous absence of the newsletter, congresses and a journal, was the incapacity of convening a General Assembly. The statutes required an Assembly every three years, but it took seven years before the

145 Letter of 31.1.1967 from Ranke to Peeters (PEETERS 8).

first General Assembly convened in Paris, during SIEF's first congress in 1971 – a congress that bore the title of 'European ethnology'. Mihai Pop was elected president – and SIEF fell into a new somnolent period of eleven years, to wake up again in 1982 for its second congress. The only regular activity that went on seems to have been in some of the commissions.

Conclusion

It is a paradox that SIEF under Peters' presidency had many traits in common with the CIAP of the days of Christiansen. There were certainly more administrative meetings in SIEF, but the output – seen from the scientific community – was not much higher; four newsletters in seven years was the most visible result. SIEF ended up in the UNESCO system exactly where critics like Erixon, Dias, Bratanić, Csermak, Rivière, Steinitz and others had wanted to see it – affiliated to the anthropological sciences. Its first congress bore the name of 'European ethnology' – a name which the latter had proposed for the organization and 'The Four' had argued so strongly against. SIEF lost its most central commission in the period, the atlas commission, and it failed in most of its scientific endeavours; the envisaged cycle of congresses was never realised, nor the publication of a journal. These tasks, however, were fairly successfully realised by those who were squeezed out during the transition from CIAP to SIEF.

The two presidents, Christiansen and Peeters, played only supporting roles. In addition, Christiansen was also assigned a scapegoat role, a role that was partly deserved. But 'the Four' were wide off the mark in their criticism and attacks, and their lack of historical knowledge about CIAP – except perhaps for Wildhaber – was striking. In their campaigning, which – to borrow a characteristic used by de Rohan-Csermak – reminded more of complicity than of preparing a report on the organization of a scientific organization, there were strong tendencies to take a stand *against* 'the Scandinavian hegemony' and material culture studies (which they more or less equated), and *for* one of the constituent parts of the discipline (folklore), as well as a conspicuous lack of modesty and moderation; all through the campaign three of them envisaged central positions for themselves. Furthermore, it is remarkable that two persons without academic positions or degrees were assigned the reformer's role in this scientific organization – a fact that did not pass unnoticed. It is tempting to regard the three as straw men, who accidentally remained in positions that were not intended for them. Actually, Ranke and Wildhaber complained more than once about the result of the Executive Board elections in Athens. The team that was elected to lead SIEF did by no means meet the sky-high expectations which they had created themselves.

The main roles in the transition act were played by Erixon and Ranke, both great scholars and leading personalities in European Ethnology/*Volkskunde*. In 1964 Erixon was the loser and Ranke the winner. A few years later it was turned around; Erixon came out as the winner. The biggest paradox in this story is probably the fact that the two did not cooperate, as their positions initially were quite concurrent. There was more that separated Christiansen and Erixon than Erixon and Ranke.

The only thing that divided the latter two was the membership question, a problem that was not insolvable. But Ranke evaded Erixon's contact efforts and rejected his compromise proposal. From 1962 to 1964, Ranke directed the campaign of 'The Four', addressed mainly against Erixon and 'the Scandinavians'.

In addition to being an internationally acknowledged scholar, Ranke was also an eminent organizer. The scientific journal *Fabula* and the *Encyclopädie des Märchens* were his work, and he was the central organiser and long-time leader of the successful *International Society for Folk Narrative Research.* What were his motives for leading a salvage operation of a moribund organization like CIAP, where he was not even member? It may well be that Ranke, as an internationally oriented researcher, saw the utility of a strong international organization and wanted to assist the rescue operation. But then a pressing question remains: Why was he so dismissive towards Erixon, and why did he not lead the operation openly? If 'the end justifies the means', what was Ranke's end or motive? With his unquestionable authority – and with some willingness to cooperate with another of the most respected (and successful) European scholars – he might probably have led CIAP/SIEF to a much better future. As it were, Ranke might despair – because he himself had contributed strongly, and probably more than anyone else, to SIEF's lack of success.

Acknowledgement

All translations to English from Norwegian, Danish, Swedish, French and German are by Bjarne Rogan.

Bibliography

Actes du congrès International d'Ethnologie Régionale (1955) Arnhem.
Actes du premier congrès international d'ethnologie européenne. Paris, 24 au 28 Août (1971). Paris (1973).
Bratanić, B. (1965) 'Forschungsbreicht. Bericht über die Tätigkeit der Ständigen internationalen Atlaskommission in den Jahren 1954–1964', *Zeitschrift für Volkskunde*, 61, 243–47.
—— (1967) 'Internationale Arbeitskonferenz über die ethnologische Kartographie. Zagreb, 8–10 Februar 1966', in *Ethnologia Europaea*, 1, 75–7.
Brednich, R. (1977–78) 'Der internationale volkskundliche Bibliographie', in *Ethnologia Europaea*, 10(2), 184–91.
—— (1977/78) *Ethnologia Europaea* Die Internationale Volkskundliche.
Bødker, L. (1965) *Folk Literature* (Germanic). *International Dictionary of Regional European Ethnology and Folklore*. Copenhagen: Rosenkilde and Bagger.
Christiansen, R.Th. (1955) 'A Folklorist's Plea for Co-operation', in *Laos III*, 9–17.
Conférence de Namur. 7 au 12 Septembre (1953) *Commission Internationale des Arts et Traditions Populaires* (C.I.A.P.) Namur.
Enzyklopädie des Märchens. Handwörterbuch zur historischen und vergleichenden Erzählforschung. Berlin-New York: Walter de Gruyter.
Erixon, S. (1951) 'Ethnologie régionale ou folklore', in *Laos I*, 9–19.

—— (1955–56) 'Internationelt samarbete', *Folkliv*, 1955–56, 141–45.
—— (ed.) (1956) International Congress of European and Western Ethnology, Stockholm 1951. Stockholm.
—— (1967) 'European Ethnology in our Time', *Ethnologia Europaea*, 1, 3–11.
Hultkrantz, Å. (1960) *International Dictionary of Regional European Ethnology and Folklore, I General Ethnological Concepts*. Copenhague: Rosenkilde and Bagger.
Laos. Etudes comparées de folklore ou d'ethnologie régionale (1951–55). Stockholm: Almqvist & Wiksell.
Niederer, A. (1983) 'Robert Wildhaber (1902–1982)', in *Ethnologia Europaea*, 13 (2), 234–36.
Rogan, B. (2007) 'Folk Art and Politics in Inter-War Europe: An Early Debate on Applied Ethnology', *Folk Life*, 45, 7–23.
—— (2008a) 'From Rivals to Partners on the Inter-War European Scene. Sigurd Erixon, Georges Henri Rivière and the International Debate on European Ethnology in the 1930s', *Arv. Nordic Yearbook of folklore*, 64, 275–324.
—— (2008b) 'Une alliance fragile', in *Ethnologie Française*, 4 (in press).
Rohan-Csermak, G. de (1967) 'Conférence internationale d'ethnologie européenne. Hässelby, 4–6 septembre 1965', in *Ethnologia Europaea*, 1, 59–74.
SIEF-Informations. Nos 1–4 (1964–1970)
Top, S. (2001) 'Peeters, Karel Constant', *Enzyklopädie des Märchens*, B (10), 685–88.
Uther, H.-J. (2004) 'Kurt Ranke', *Enzyklopädie des Märchens*, B (11), 207–14.

Archival Sources

(Abbreviations used in the footnotes)
Delargy Center for Irish Folklore and the National Folklore Collection, Dublin (DUBLIN)
Institut für europäische Ethnologie, Georg-August-Universität, Göttingen (GÖTTINGEN)
Landsmålsarkivet, ULMA, Uppsala. (SOFI, ULMA)
Meertens Instituut, Amsterdam (MEERTENS)
Musée des Arts et Traditions Populaires, Paris. (MNATP)
Special collection under CIAP/SIEF: Documents from Peeters, Antwerp (PEETERS)
Museum Nacional de Etnologia, Lisbon (LISBON)
Special collection: Jorge Dias' archives (DIAS)
Nasjonalbiblioteket, Oslo. Håndskriftsamlingen (OSLO)
Nordiska Museet, Stockholm. (NORDISKA)
Special collection: Sigurd Erixons samlingar (SE)
UNESCO/The League of Nations, Paris (UNESCO)
Universitetsbiblioteket, Lund. Handskriftsamlingen (LUND)
Österreichisches Museum für Volkskunde, Vienna (VIENNA)

Chapter 3

Small National Ethnologies and Supranational Empires: The Case of the Habsburg Monarchy

Bojan Baskar

The expression 'small national ethnologies' suggests a central and east-European provenance. In a West-European context, the meaning of the term, although 'ethnology' is replaced with 'anthropology', is generally not self-evident. The self-evident bond between small national ethnologies and the central and east European space is further derived from the widely held conviction that the central and east European ethnologies are nothing other than an expression of romantic national movements and small-nation statisms. As an instrument of nationalism, they are held to be predetermined to exhaust themselves in inventing (if not forging) a national culture, codifying its corpuses, sublimating dialects into literary languages, determining the boundaries of ethnic territories, etc. For this reason, they are not scientific. At times, it is also presumed that ethnologists from that part of Europe lack any theoretical background, belong to no intellectual traditions; they may be imagined as sheer anti-intellectual populists. It is further presumed that they are interested only in their own *gemeinschaft*, therefore indifferent to any other culture. Such presumptions make the expression 'small national ethnologies' suspicious a priori and broaching this subject is liable to encounter resistance. Indeed, who would enjoy listening to the Ruritanians talking to the Bragadocians, now that they have all come into possession of their own mini-states, and are no longer happy with their 'village greens'?

To a certain extent, these notions about small ethnologies can be extended to their west, and particularly north, European counterparts. Yet there is an important difference in the general outline of the nation-building between the west and the centre and east. It has been described and conceptualised in a variety of ways. Perhaps the most important of difference is to be searched for in contexts other than those of the nation-building process. In Central and Eastern Europe, national movements emerged in the framework of absolutist multi-ethnic empires whereas in Western Europe the nation was often formed within the pre-existent 'national' state. The implications of the central and east European situation, where the national movements were embedded in multinational dynastic empires (the Habsburg, the Ottoman and the Romanov), have not been sufficiently studied. They are as a rule more complex than the national movements embedded in a framework of a national state. Numerous social scientists, not used to thinking in terms of multinational empires, are prone to overlook this circumstance and to take for granted the framework of the nation-state: such has been the case of Parsons and Durkheim who mixed up social system and society with the nation-state (Strassoldo 1991: 176). Thus, this framework can easily, and without being noticed, become the 'social framework', i.e. the social context of national movements.

Empire-building *vs.* Nation-building?

After the Second World War, with the political division of Europe into the West and East, a tendency emerged to increasingly rely on dichotomies and binary oppositions as tools of conceptualizing the division. Binary oppositions are easily reified and therefore of dubious assistance in the analysis. The dichotomy of the empire-building and the nation-building, proposed by the prominent historian of anthropology George Stocking, Jr., is one of the more surprising – and also challenging:

> In these terms we may contrast anthropologies of nation-building and empire-building, and within the former we may distinguish between those continental European traditions where strong traditions of *volkskunde* focussed on the internal peasant others who composed the nation, or the potential nations within an imperial state, and those postcolonial nation-building anthropologies of the Third World, whose relation to internal otherness in some cases approximates that of an internal colonialism. (Stocking, Jr. 1984: 5)

Here, the nation (nation-state) and empire are contrasted, but they may also be opposed if they are considered in terms of *mastering* – which is a classic distinction from political philosophy: mastering oneself as against mastering others. Such an Aristotelian opposition would suggest that colonialist nations developed the empire-building ethnology (anthropology), without simultaneously developing also the nation-building ethnology. This might be the case for Britain whereas it certainly does not hold for France. Besides an ethnology of exotic societies, the French have always been taking care of their folklore, eventually renamed in the ethnology of France. The discipline has to be conceived of as a considerable component of the nation-building process inasmuch as the notion of nation-building does not indicate only that phase of the nation-formation which ends with its achieving the attributes of a fully fledged nation, but refers to the continuous reproduction of national identities in the ever changing conditions. A deep involvement of the ethnology of France in the process of reproduction of national identities can be discerned, for instance, in its frequent exploitation of the *topos* of diversity-in-unity. Yet France is not really an exception among the postcolonialist nations. In the Portuguese case, despite the existence of an empire and the absence of a national problem in the classical sense, the founding fathers of anthropology also developed a nation-building anthropology that was searching for Portuguese national identity (Leal 1999: 179). The exception that calls for explanation is rather the British anthropology.

There is a distinction between nations with their own state who are also in possession of an empire, and (potential) nations without their own state and embedded in a multinational empire. Within this frame of reference the latter are nations that are formally in the same position as potential nations in the colonial empires owned by the nation-states from the first group. The difference between them lies primarily in the kind of empire, as there are overseas colonialist empires built by powerful European nations, but also, as in the first case, the traditional (dynastic) European, and partly Asian, empires that had emerged before the era of the nation-states. The overseas empires are discontinuous whereas traditional empires are contiguous; the demise of the first and the second had rather different consequences for the national identity of a dominant national group (cf. Barkey and von Hagen 1997, Geiss 1994). The contrast of empire-building and nation-building may thus be translated in the contrast of two types of empire. Naturally, this contrast should not be dichotomised: when the Austro-Hungarian monarchy for instance was given a mandate over Bosnia-Herzegovina, its policy showed clear colonial features, comparable to those of the West-European colonialist nations in their overseas territories, although the two countries had previously bordered to one another and for this reason Austro-Hungarian monarchy after the occupation of Bosnia-Herzegovina remained a contiguous empire. It also should not be overlooked that some West-European colonialist states acquired part of their colonial possessions in their neighbourhood, particularly on the opposite shore of the Mediterranean Sea such as Spain in Morocco, France in Algeria, Tunisia and Morocco, and Italy in Libya. Such quasi-contiguous forms represent an intermediate form of empire. Perhaps the most unusual case had been Venice whose Eastern-Adriatic possessions started in Istria, only one hundred Venetian miles away from the metropolis.

While imperial colonialist nations are not necessarily free from caring for their own national identity (i.e. nation-building), national movements within multinational empires are as a rule not limited to the nation-building since they participate in the reproduction of the totality of empire (i.e. empire-building). Thus, the West and the Central-East European ethnology and anthropology have been playing both the nation-building and the empire-building roles.

When Nation-Building is Empire-Building

The assumption that nation-building and empire-building are incompatible is related to an erroneous conception of the nature of national movements in multinational empires. An impetus, intrinsic to these movements, is assumed which drives them, from the very beginning, to demand full political independence and their own national state. Nationalist historiography, exalting the 'millennial yearning of the nation for its own state', is plainly interested in maintaining such a conception. Less obvious is why certain highly influential liberal theories of nationalism prefer to share the same view, and correspondingly picture the national movements of small nations, these 'latecomers to the historical scene', as a raving action of virulent ethnics, populist demagogues or partisans of closed systems. Such a notion was partly refuted by Miroslav Hroch who, on the basis of a careful comparative analysis, proposed a three-stage periodization of the national movements in Central and Eastern Europe (Hroch 1985, Hroch 1995). Demands for self-determination or full independence as a rule emerged at a very late stage. With the collapse of empires following the First World War, the majority of national groups who were granted independence gained something they had not demanded in their political programmes (see also Roshwald

2001). As Aviel Roshwald has rightly warned (2001: 2) we should bear in mind that this kind of ternary developmental typology runs the risk of lapsing into a historical determinism that is in some ways analogous to the teleological mentality pervading nationalist ideologies.

Concerning Hroch's Phase A of national movements (scholars discovering and studying folk culture), it can be asserted that the nation-building virtually entirely coincides with the empire-building insofar as the empire recognises and admits its diversity of peoples and cultures. Pietro Kandler, a Triestine local historian from the second half oh the nineteenth century, was a textbook case of a scholar from the Phase A. He strolled the whole Istrian peninsula, searching for inscriptions, coins, tombs, tumuli, Roman roads and other antiquities, preferably Roman ones. Not really untypical of a scholar of this kind was his interest in other ethnicities in the region; thus he regretted his inability to understand the 'Illyrian' idiom (De Franceschi 1989 [1926]). When Italy united, the Italian national movement in the Austrian Littoral entered the Phase B; simultaneously irredentism emerged within it, yearning after the annexation of the region by the new Italian state. Istrian irredentist scholars admired Kandler as their teacher, yet strongly resented his unshakeable conviction that for Trieste and Istria it was wiser to remain part of the Austrian monarchy. Kandler clearly distinguished between the 'genetic' (i.e. cultural) and the political nation. In Trieste, despite the escalation of the ethnic conflict between the Italians and the Slovenians, irredentism remained weak until the First Word War. Neither did Yugoslavist ideas resound among the Slovenian group. Regardless of their nationality, also those workers who were organised in a strong Social Democratic party preferred the Austro-Marxist variant which advocated reforms within the framework of the existing state (Cattaruzza 1993, Vivante 1997[1912]). Thus, adherence to cultural nationalism – and the corresponding renouncement to political nationalism – was far from being exceptional. Also in the Phase B and C, the overlapping between nation-building and empire-building remained significant.

A variety of ethnographic activity aimed at establishing a national culture is liable to be consonant with the affirmation of the empire if the latter prides itself on the benevolence toward the peoples it contains and the diversity of their languages and customs. Furthermore, it should be recalled that actors of national movements belonged to several referential groups, and therefore they held several identities. They mastered different roles that they played at different occasions. On working days, they would wear their ethnic costume; on Sunday, though, they might put on the coat of the supranational patriotism it. And vice versa: cosmopolitan personalities could occasionally assume the role of the patron of a local cultural nationalist organization. Such was, among others, the case of the well-travelled writer and ethnographer Mara Tschopp Berks who wrote exotic prose about the Balkans, South Slav women and Gypsies, held a literary salon in Vienna, lived an unconventional life, was a member of the Gypsy Lore Society and collaborator of the Hungarian Ethnographic Society. This notwithstanding, she accepted to be the patron of the local Slovenian nationalist *Sokol* society in Celje where her husband possessed a castle (Grdina 1999). Perhaps all of the monarchy's national groups had poets who sang about misfortunes and hopes of their nation, yet composed an ode to the Kaiser when he came to visit.

When the Hungarian Ethnographic Society was constituted in 1889, its agenda was the following: 1) strengthening 'imperial patriotism' through the mutual knowledge and familiarizing of imperial peoples; 2) studying comparatively the peoples of the Hungarian state contained within the empire; 3) consolidating Hungarian national identity (Kósa 2001: 1–2). It would be simplistic to consider the first and the second aim of the Society a sheer concession to the regime and to hold only the third aim as the true one. In practice, much more energy and time was quite probably sacrificed to the building of national identity than the other two. Still, both representatives of nations and representatives of the empire could agree that imperial and national patriotism were perfectly compatible even within the same institution. Choosing crown prince Rudolph as the Society's patron was naturally not a coincidence as he was an apparent promoter and sponsor of imperial ethnography, conceived as a tool to foster the supranational patriotism and love for the Monarchy. (For his sponsoring of a big ethnographic project which resulted in twenty four volumes of *Die österreichisch-ungarische Monarchie in Wort und Bild*, see Fikfak and Johler 1998, Stagl 1998.)

The Small and the Big Homeland

The linking of two *topoi* – diversity-in-unity ('sub pluribus unum') and the small and big homeland – proved to be an effective mechanism for harmonizing national identities with the supranational Austrian identity. The articulation of the national and supranational level can be compared to the articulation of the regional and national level in a national state. At the same time, the relationship between the regional and national level in the national state can also be compared to the relation between the regional and supranational level in the multinational empire. The latter offers belonging and homeland; in addition to the local, regional (*landespatriotismus*) and national patriotism, it proposes a supranational patriotism. It thus seems at first sight that there is no difference between the empire and the nation-state, except in the number of attachments to homelands of varying level and size. Yet, in the case of the empire, the difference between the regional and national level is relative, whereas it is, so to speak, absolute in the case of the nation-state. Moreover, the ultimate criterion of imperial supranational patriotism is the emperor himself. This peculiarity of dynastic patriotism was expressed succinctly by Franz Josef in person, who, after somebody was recommended to him as an Austrian patriot, retorted: 'He may be a patriot for Austria, but the question is whether he is a patriot for me' (cited in Mann 1995: 51).

The use of this mechanism in imperial Habsburg ideology calls for a comparison with a similar mechanism used in French national ideology, though France as a strongly centralised and assimilationist nation-state appears to be the opposite of the Habsburg Empire. In the case France's nation-building, this mechanism was studied in detail by the French ethnologist Anne-Marie Thiesse (Thiesse 1991, Thiesse 1997). In addition, the heyday of the two *topoi* seems to have been roughly synchronous. Thiesse has shown that the 'discovery' of French diversity gained momentum when France was defeated in the Franco-Prussian war of 1871. The humiliating Austrian defeat at Sadowa in 1866 by the same power was crucial for the development of comparable Habsburg ideology of diversity of the imperial peoples and plurality of their small homelands, as it was followed by the expulsion of Austria from the German Reich. In response the oriental mission of Austria was invented. The empire started to legitimate itself as the protector of the small nationalities of South-East Europe (Agnelli 1971).

Is the Sovereign State a Necessary Prerequisite of Colonialism? Croatian 'Small Colonialism' within the Habsburg Empire

Ethnicities within empires may be potential nations aspiring after their own states, but they may also be potential colonialists (or 'imperialists') acting within empires on their own behalf. The Croatian national movement of the mid-nineteenth century is a case in point. This movement consisted of two principal currents, both of which showed a peculiar interest in Bosnia-Herzegovina. The first, Illyrianism, emerged in the 1830s, and it had a generally hostile attitude towards the Ottoman state. Within this movement, there were programmes ranging from the unification of all of the Southern Slavs ('Illyrians') in one state (the maximal programme) to the unification of certain South Slavic peoples (Bosnians included) in the frame of the Habsburg monarchy (the minimal programme). At the start, the leader of the movement, Ljudevit Gaj, and some other members indulged in conspiratorial activities aimed at raising a rebellion of South Slavic peoples in Bosnia-Herzegovina, secretly trying to convince the Russians to help liberate the Bosnian brothers. Ante Starčević, a renegade of the Illyrian movement, founded a new movement which was above all hostile toward the Austrian monarchy. The Ottoman state in particular and Islam in general, on the contrary, had unusual sympathies from his part. While Gaj wanted to destroy the Ottoman state, Starčević wanted to destroy the Habsburg state, and both wished to unite the Bosnians with Croatians either within the Austrian empire or in independent (and Greater) Croatia.

The tendency towards the unity of Croatians (in Illyrianism still appearing under three different regional names, i.e. Croatians, Slavonians and Dalmatians) with Bosnians, shared by the two varieties of Croatian national movement, invariably implied a new territorial cutting (eventually to be called the Croatian ethnic space). The independent and greater Croatian state should have been carved out from the two empires. In the movement led by Starčević, such a state was undoubtedly conceived as a nation-state; due to the internal ethno-religious boundary between Catholic and Muslim 'Croatians', however, the Muslim part of the state would, in many regards, look like a colonial achievement of the 'core' Croatia. On the other hand, snatching Bosnia-Herzegovina from the Ottoman state, and its annexation to the Austrian empire (with a possible union of Bosnia-Herzegovina and Croatia as an autonomous land within its frame) would have been a classic move of imperial expansion. Yet the aspiration for annexation does not necessarily emerge in the imperial core or among the *staatsvolk*. In the case of the Illyrianist movement, it emerged in the plain periphery, in an imperial borderland where such an annexation was conceived by the local national movement simultaneously as an imperial expansion (on its own behalf) and a formation of its own national state.

Such an ambiguous mixture of empire-building and nation-building can also be observed in the Croatian interest in geography, the history and ethnography of Bosnia-Herzegovina, which developed in the same period. In the mid-nineteenth century, the Croatians still lacked professional ethnologists, thus the ethnographic description of Bosnia-Herzegovina was primarily the domain of travel writers. While travels of early Illyrianists were often motivated by their conspiratorial activities, and their descriptions consequently tendentious in informing about affliction of the Christians in Bosnia, increasingly elaborate descriptions of Bosnia emerged after 1850.

In 1858, an outstanding description by a deputy of the parliament appeared. Ivan Kukuljević, crossed the border free of conspiratorial tasks (Kukuljević Sakcinski 1997, Grbić 1997), and thus established a tradition of observers of Bosnia-Herzegovina who developed a predominant interest in Bosnian Muslims rather than Bosnian Catholics (the future Bosnian Croatians). Though the latter were often their hosts and privileged informers, in the eyes of travellers and ethnographers they were clearly less interesting. To Kukuljević, who had long yearned to see Bosnia, the journey was a deep emotional satisfaction. He structured his travel experience in the binary terms of the familiar and the alien, the close and the distant, the Other and the Self. The ambivalence of familiar and unfamiliar, imperial and national, is peculiarly condensed in the description of entering Bosnia at the border on Una, a 'tiny river' which 'divides into two immense empires and one single nation' (1997: 328). The perception of a clear cultural boundary, discernible as a waste landscape on the other side, a difference of costume, a view of an apathetic squatting Turk, is juxtaposed to a feeling and evocation of common ethnicity betrayed by the melancholic 'oi!' of an antique folk song, which 'can be heard from both banks of the Sava river, magically uniting the spirit of two related bodies' (1997: 328). Under the Islamic-Turkish surface, Kukuljević's love for his nation is able to detect a Slavonic and Croatian substratum: it shows itself in the cheerful and jocose character of the people, in the purity of the Bosnian tongue, love of folk customs, oral tradition which keeps the memory of a common ethnicity alive. More satisfactory than discovering the shared national substance, though, is a recognition of figures of the Other, known to him from his previous reading: the apathetic squatting Turk above the river, the Turkish meal without cutlery, and especially the listening to the muezzin's evening singing which he finds superior to the singing of Christian bells, as regards the intensity of emotion excited by it (1997: 337). In this travelogue one finds a sequence of Orientalist *topoi*: towns that appear beautiful only when seen from afar; the ubiquitous stagnation and fatalism; negative attitude toward agriculture; etc. For this reason, the contention that Kukuljević went to Bosnia in search of the familiar, but found there the unfamiliar, seems somewhat simplified. Undoubtedly, Kukuljević's travel description has a 'nation-awakening' intent, yet it is simultaneously a proto-ethnography of an Oriental Other from an Asiatic empire.

This 'external' encounter with a cultural boundary and Oriental Other was transformed in the Starčević version of Croatian national ideology in the 'intranational' cultural boundary. Consequently, the Oriental Other became a half-familiar, half-unfamiliar Croatian: the Muslim Croatian. This ideology was ottomanophile and islamophile; and as such it represented an unprecedented innovation. As the editor of the almanac for the year 1858, Starčević had tried to have the Croatian public familiarised with Islam. He described the Prophet's life and the Qur'an as historical phenomena rather than embodiment of the Evil (Gross 1973: 33). In his booklet dealing with the Oriental question (its first version was written in 1876) he asserted that the Ottoman empire (contrasted favourably to Austrian and Russian empires as well as Serbia) was a highly civilised, developed and freedom-enjoying country which could well serve as the exemplar of religious tolerance. Bosnian Muslims were declared both the oldest nobility of Europe and the Muslim Croatians (Starčević 1992 [1898]). Bearing in mind his islamophilia, it is logical that rather than contemplating their (re)catholicization he proposed a new idea of Croatian nation as composed of two religions, Catholic and Muslim. The political process of nation-formation had to acknowledge that not all Croatians were Catholics. Whereas the 'Great Serbian' national movement envisaged a 'reduction' of Bosnian Muslims back to their 'original Orthodox faith', the 'Great Croatian' nationalism of the Starčević's Party of Right was more sophisticated and for this reason had more success with converting Muslims into Croatians.

After the Austrian occupation of Bosnia-Herzegovina (1878), Croatian literature experienced a tide of historical novels and short stories that idealised, glorified – and orientalised – the Bosnian Muslims. The authors of these works who often were followers of the Starčević's movement, though they could also adhere to other currents of the Great Croatian nationalism, were fond of assuming Bosnian Muslim pen names (such as Osman-beg Štafić, Ferid Maglajlić, Omer and Ivo, Jusuf, Ibni Mostari, writers in tandem Osman-Aziz) (Banac 1984, Hadžijahić 1990, Rizvić 1994). These works aimed at promoting the Croatian identity among Bosnian Muslims as well as having the Croatian public familiarise with the latter.

The Croatian academic ethnology emerged in the 1890s. Its founder, Antun Radić, showed a vivid interest in Bosnia-Herzegovina. In 1899, he undertook an ethnographic journey there. The account of his travel shows a pronounced methodological concern. Its principal interest lies, however, in the observed distinction of two cultures (the lordly and the peasant) and the absence of the first among Bosnian Muslims. Implications of this observation of a non-existent distance between the two for the theory of Orientalism are notable (A. Radić [1936 1899]).

For the period of the Austrian mandate in Bosnia-Herzegovina it is still easier to show that the Croatian ethnographic interest in Bosnia-Herzegovina was not only 'nationalist' but also 'imperialist' or 'colonialist'. The Austrian administration expectedly stimulated colonization. It is assessed that about 200,000 settlers from almost all parts of the empire moved in Bosnia (Kržišnik-Bukić 1996), the neighbouring Croatians having formed the most numerous single ethnic group. Political endeavours related to Bosnia by the Antun Radić's brother Stjepan were to a significant extent bound to this colonization. Stjepan was later to become one of the top politicians of the first Yugoslavia, and together with his brother, the founder and the ideologist of the Croatian peasant movement (which became the strongest Croatian political party). In his book *Modern colonization and the Slavs* he recommended to the Hungarians and the Danube Slavs (primarily the Croatians, Czechs and Slovenians) to make good use of religious connections of the Muslim Slavs in Bosnia-Herzegovina in order to turn the Near East in their colonial domain (Stoianovich 1994: 297) – naturally within the framework of the Austro-Hungarian monarchy and its dream of *Drang nach Osten*. In his booklet *The Living Croatian Right to Bosnia and Herzegovina* (1908), on the other hand, he argued for the special European, Habsburg, and Croatian right to Bosnia-Herzegovina (S. Radić 1993 [1908]).

Who Studies Whom? The 'Inequality of Power' Principle

Although contemporary social sciences are satiated with discourses on power and the inequality of power, there are disciplines that do not talk about power. In ethnology, and especially in social anthropology, the silence about the inequality of power when unequal distribution of field research sites is commented upon, is particularly striking. While one hears the frequently voiced, yet seldom-questioned (but see Bernard and Digard 1986) opinion among 'Western' anthropologists that the tide of 'anthropology at home' is due to increasingly difficult accessibility of 'exotic' fields (i.e. to a specific anthropologists' *lack of power* to leave for the 'real' fields), it is generally assumed that the preoccupation of East-European ethnographers and folklorists with the folk culture is due to a *lack of interest*, lack of cosmopolitan curiosity in or indifference to the Other.

The tendency to choose the natives among whom one feels safe seems common. It hardly happens that an ethnologist chooses natives that would look down on him/her, enjoy higher standing, be richer and more powerful, and couch assumptions about the inferiority of his/her culture. An African ethnologist who finds his way in Austrian vineyards or learns to drive the car during his fieldwork in the USA stages his exceptionality in a peculiarly picturesque and unthreatening way. Likewise, he is exposed to a benevolent curiosity towards the natives who normally do not expect any scholarly expertise of him. Moreover, the ethnological study of nobilities, for instance, is quite exceptional, especially when it is based on fieldwork. Notwithstanding this, the fact remains that for ethnologists the aristocratic ways are as interesting and important (and 'exotic') as the peasant ways.

Besides the hierarchy of social classes and their respective cultures, ethnologists have to do also with a hierarchy of national cultures. The latter accounts for a seemingly odd detail that the central and east European ethnologists do their field research more frequently in 'exotic' societies (which is an infrequent case) than in a neighbouring society that has traditionally dominated their own. A predicament is thus being perpetuated which fully developed already in the nineteenth century, if not earlier. It is difficult to imagine, in the Austro-Hungarian empire, the Czechs developing the ethnology of the Germans, the Slovenians that of the Italians, the Ruthenians that of the Poles, the Slovaks that of the Hungarians, and the Bosnian Muslims that of the Croatians and Serbians. To the national groups who had no 'underdog', only 'their' peasants remained, and perhaps 'their' Gypsies too. On the other hand, the 'top down' choice of fieldwork location was the rule: the German speaking Austrians studied, to some extent, all national groups; the Italians were somewhat interested in the Slovenians and Croatians; the Croatians and the Serbs (later joined by the Slovenians) showed interest in the Muslims of Bosnia. This principle of division of labour remained in power in socialist Yugoslavia where, despite the officially proclaimed 'equality and brotherhood' of 'nations and nationalities', three nations were discreetly 'more equal' than others: only the Serb, Croatian and Slovenian ethnologists somewhat dared to go beyond the limits of their 'ethnic territories' within Yugoslavia. In Soviet Union, where there was only one *staatsvolk*, only Russian ethnologists were crossing the boundaries of their ethnic culture while ethnologists from all other national groups preferred to study their own people (Dragadze 1987: 160-61).

In a multinational state, its *staatsvolk* is not necessarily privileged, at least not in every aspect, though it is induced, by its very role of a core ethnicity, to be interested in others and to study them: perhaps aspiring to assimilate them or, following the principle of *divide et impera*, fomenting discontent among national groups and weakening those who are too strong with the promotion of new nationalities. The role of the *staatsvolk* also obliges the core nation to articulate relations among other national groups composing the whole complex, to encourage the display of their folklore traditions, thus granting recognition to the cultural diversity of the whole. Besides core nation, there are intermediary national groups, politically (and sometimes also culturally) subordinated to the core nation, though dominating, often in an unrelenting manner, weaker national groups, or groups at an earlier stage of nation-building, in their neighbourhood. As a rule, they strive to assimilate them to their own 'superior' culture or to induce them to believe in their mutual ethnic belonging.

The assumption, characteristic of British social anthropology, that east European ethnologies are closed systems breeding a nativist lack of concern for the rest of the world, is efficiently called into question by those UK-educated but non-British anthropologists who, despite their 'cosmopolitan' training, conducted their field research 'at home', e.g. in various African countries, India, Spain, Greece, Cyprus. If inequality of power is not taken into consideration, one may easily lapse into essentialist explanations, based on assumptions about 'anthropological', i.e. 'deeply rooted', characters of certain peoples and certain regions. It can be agreed that the persistence of the *volkskunde* research traditions in contemporary central and east Europe is 'per se such an unusual phenomenon that it would deserve an anthropological analysis in the full sense of the term' (Hann 1990: 261), though only on condition that also the tradition of British social anthropology, due to its peculiarities, deserves the same analysis. Cultural peculiarities of central and east Europe cannot be seriously studied in isolation from the cultural peculiarities of west Europe.

Literature and Ethnology

The role and significance of literature in the formation of national identities in central and east Europe is well known. There, the 'anthropological' relevance of literature has often been brought to light by other disciplines, while ethnologists remain reluctant to make literary agencies their research subject. As Anne-Marie Thiesse suggests in relation to French ethnology, this may be due to the similar development of both folklore and ethnographic scholarship, and the rise of regionalist literature.

Folk traditions and costumes were the subject of description by folklorists and regionalist writers alike; their interrelations were both of alliance and competition (Thiesse 1991: 12). However, the contemporary ethnology of France and a certain history of mentalities turn a deaf ear to the fact that the regionalist literature forms a part of their own genealogy (Thiesse, 1991: 289). Consequently, the cause of the mentioned reticence might be the fear that, by exposing the genealogical context of folklore and ethnography, a doubt as to their scientific status would emerge.

The thriving of regionalist and localist literatures in the second half of the nineteenth century was a wider European phenomenon. In the multinational Austrian empire, regionalist literature defined itself against national as well as so called supranational literature. The affirmation of a local/regional identity often went hand in hand with the affirmation of the supranational Franzjosephine world, in the first place in multilingual and multicultural regions of the empire. Yet the idyllic affinity of the regional and the supranational was sometimes disturbed, as regional and national identities were liable to overlap.

In the field of literary science, Claudio Magris studied this circumstance in detail decades ago. He conceived it in relation to the literature of the Habsburg myth and characterised the solidarity of the local/regional and the supranational as a 'harmony of the supranational unity and the local savour' (Magris 1963: 177). In this literature, the interest in the local and the regional is to a significant extent materialised in the process of description of the folklore of local ethnic communities and their 'intercultural' interactions. Several nationally mixed and often borderland regions with their provincial capitals have thus become famous in literary history: Vienna, Prague, Czernowitz (in former Bukowina), and Trieste. In all of these cases, the central role had been played by Jewish writers who, besides German writers, showed more interest than any other in the literary ethnographic description of co-residing national groups and their interactions. It may appear at first glance that the Jewish writers were exempted from the 'law of the inequality of power'. As a discriminated-against and often persecuted ethnicity, the Jews 1) started to play the role of the 'observers of the others' only after citizen rights (including the possibility of assimilation into the dominant nation) had been granted upon them; and 2) mostly choose the language of the dominant national group or a culturally prestigious language (i.e. the German or the Italian in the Triestine case). The major part of the Austro-Hungarian ethnographic literature was written in German: the phrase 'ethnographic short story' was then introduced to describe the work of Bukowinan writer Karl Emil Franzos (Moraldo 2002: 11), one of the most salient representatives of this literature. Yet the extent and the interest of homologous literature in other languages should not be underrated. This literature is less known since in the national literary histories of respective nations it is either suppressed or reinterpreted as a nation-building literature. A case in point is Mór Jókai, the most prominent and very popular Hungarian prose writer of the nineteenth century, who wrote nation-building as well as supranational novels, the latter having had the Ottoman empire as their setting.[1]

---

1 Contemporary and later critics have generally praised Jókai's historical novels in which there are Hungarian national themes or a combination of these and Oriental elements. His novels in which the Orient dominates have, however, been either criticised as badly composed

Compared to academic ethnography, ethnographic literature is an easier way of proceeding. Academic ethnography can flourish only where the whole academic sphere is relatively developed, where long professional training is provided, where there are research institutions available, together with the system of their financing, etc. The literary production, on the contrary, is incomparably more modest; the conditions needed for its presence were fulfilled also within regional frames. This largely explains why (proto)ethnology first emerged in literature as much as why non-academic literary ethnography (or ethnographic literature) persisted as a predominant, if not the only option in less developed, less powerful, less enterprising and poorer national groups. Literature is an easier way also because it usually does not require a field interaction with actors from other linguistic and cultural groups.

A case in point is the Croatian colonial interest in Bosnia-Herzegovina. Within that frame, a travel literature emerged, which was followed by the literary fiction and historical literature taking place in Bosnia, while the extent of the academic ethnological scholarship, despite the initial impulse of Antun Radić, remained quite limited. Stjepan Radić's plans regarding the 'Austro-Slavic' *Drang nach Osten* were even less conclusive in establishing a would-be Croatian ethnology of the Middle East.

Thus, the relative self-containment of small national ethnologies is not simply a consequence of their nationalist ethnocentrism but also, and more importantly, a consequence of their own lack of power and resources as well as their academic underdevelopment which does not enable ethnologists to conduct systematic and continuous research outside their national territory. When self-containment becomes an institutionalised practice, it naturally tends to persist also in those circumstances where some favourable prospects of an outward expansion emerge. Thus, for instance, in socialist Yugoslavia, where a few more important universities became developed enough to proceed with an outward orientation. The reputation enjoyed by Marshal Tito's Yugoslavia in most of the non-aligned countries could make it easier for Yugoslav ethnologists to launch their research in the Third World. Political support for such an enterprise could quite possibly be negotiated, especially on the level of federal instances. Yet this expansion was never contemplated, and the inertia of dealing with folk (and working class) culture was seemingly most responsible for this missed opportunity. Not surprisingly, the failure to inaugurate an ethnography of non-aligned countries was partly compensated by a rich harvest of the travel writing literature from those countries.

The long-term limitation of the central and east European ethnographic curiosity to ethnographic literature partly explains why in this part of Europe the standing of literature in society is so much more important than in the West. That literary culture is more important, more influential and more authoritative than scientific culture is an additional aspect of less developed social sciences, and a weaker representation of ethnology in the formation of national identity in the respective societies. There, writers take it for granted that they can talk authoritatively about everything which normally falls into the domain of social sciences, and they never think of consulting

and over-romantic or simply ignored'. Cf. Bellér-Hann 1995, p. 227. The clear boundary between the nation-building and the supranational approach to interethnic interactions in the literature is naturally impossible to establish social scientists on the subject. They believe to be competent in the matter of nation and national identity, hence also in the matter dealt with by ethnology. Contrary to the widely held opinion, ethnography in central and east European nations had in general a weak impact in the nation-building process (for the marginal status of Croatian ethnology within the Croatian society, see Čapo Žmegač 1999). This is why politicians and writers, who in these societies are often would-be politicians, have been able to ignore ethnology.

Conclusion

This paper suggests that the topic of small national ethnologies has a specific relation to the central and east European space because that space has been almost entirely occupied by several multinational empires until the beginning of the twentieth century. Implications of this circumstance for processes of nation-formation in central and east Europe are huge, yet they are inadequately studied. The history of national ethnologies of this part of Europe has been deeply affected by the same circumstance. The ignorance of historical imperial contexts, where national movements of central and east European nations occurred as a rule, results in a distorted notion of the character of these movements and, consequently, of respective national ethnologies. National movements in central and east Europe in general do not conform to the descriptions offered by simplistic theories of nationalism, especially Gellner's. Being overwhelmed by one's small ethnos and its charming language proved to be not so incompatible with the pride of belonging to a big and important supranational homeland. At the same time, the role of ethnology in the making of central and east European nations may not have been as central as it was commonly held.

Small nations, of course, are not necessarily powerless, poor and non-enterprising. Some of them have been building empires (the Danes and the Portuguese), others such as the Irish and the Norwegians, were, on the contrary, constrained to develop their separate identity within larger multinational polities. A couple of years ago, a polemic between a Danish anthropologist and a specialist in Iceland, and several 'native' Icelandic anthropologists took place on the pages of *Social Anthropology*, the journal of the European Association of Social Anthropologists. Though the polemic did not turn into a dispute between anthropologists and ethnologists (all participants considered themselves social anthropologists), it revealed the fact that a certain re-enactment of colonial roles was taking place, with the Danish anthropologist claiming for herself the role of a hard scientist, while expecting the 'natives' from the remote periphery to play the role of Herderian nationalist ethnographers (which they did not).

On the other hand, there are in central and East Europe several peoples who are not small by any criteria, such as the Poles and the Ukrainians, not to mention the Russians. Yet, their story (with Russians exempted), as regards their national formation, is perfectly comparable with the story of smaller nations in their neighbourhood. Hence the size of nations and national ethnologies (defined in absolute numbers of their members) is not a factor with a significant explanatory power, except where immediate implications of small social systems are being studied.

The comparative study of national ethnological traditions making up European ethnology can gain further insights if comparison takes place on the level of imperial contexts within which national traditions are embedded. Thus, not only do we avoid a 'nationalist' approach based on the assumption that the nation-formation is endogenous (nations are made from their own substance), but also manage to grasp ethnological traditions in the all-embracing processes of their making. This also implies that virtually all European nations, actual as well as potential, were placed in imperial contexts: some as empire-builders, others as subject 'nationalities' within multinational empires (yet often participating in the building and supporting of empire), others as possessors of national states who were for various reasons not very successful in their expansionist aspirations and thus were perceived as nation-states rather than empires (e.g. Italy and Serbia). In Europe there were no nations possessing their own mono-ethnic nation-state and showing no appetite for territories in the neighbourhood or overseas colonies.

Processes that engendered ethnological or anthropological knowledge cannot be soundly separated from processes of power. Both took place in an imperial context. There was always a question of expansion and of resistance against it, of subordination and efforts to be liberated from it. Various 'constituents' of anthropological knowledge, such as social and cultural anthropology, biological anthropology, folklore, *volkskunde*, ethnology, and so forth, developed techniques of observation and description of different characteristics of various human groups or collectivities; at the same time, they played a significant role in (re)defining and naming these very collectivities. Clearly, this explains why a comparative study should not be reduced to one single constituent (e.g. ethnology as opposed to anthropology, or ethnography as opposed to folklore). An isolated comparative study of disciplines making up the whole of anthropology is bound to be unproductive since it cannot grasp the interactions between these disciplines and the logic of their articulation. Yet this articulation is extremely indicative of relations of subordination and resistance of national groups that define the imperial contexts.

It is not our intent to suggest that small national ethnologies, by their smallness or their often peripheral position, represent an epistemic advantage if compared to great ethnologies. Still, it could be inferred that different imperial contexts left their imprint on all national ethnologies originating in them. Certain common characters, affinities and specific insights are not easily accessible to ethnologies from other imperial contexts. Among them one could mention specific experience of the centre and periphery, characteristic of the heirs of the Austrian empire (whose essence, as Joseph Roth claimed, was not the centre, but the periphery), or specific experience of the co-existence of ethno-religious millet communities, common to the heirs of the Ottoman Empire. Whether they know it or not, national ethnologies are the sites of memory of empires rather than the sites of memory of national cultures.

Acknowledgements

I owe special thanks for comments, suggestions and support to Reinhard Johler, Cathie Carmichael, Jasna Čapo Žmegač and Renata Jambrešić Kirin.

Bibliography

Agnelli, A. (1971) *La genesi dell'idea di mitteleuropa.* Milano: Giuffrè Editore.

Banac, I. (1984) *The National Question in Yugoslavia: Origins, History, Politics.* Ithaca and London: Cornell University Press.

Barkey, K. and von Hagen, M. (eds) (1997) *After Empire: Multiethnic Societies and Nation-Building. The Soviet Union and the Russian, Ottoman and Habsburg Empire.* Colorado: Westview Press.

Bellér-Hann, I. (1995) 'The Turks in Nineteenth-Century Hungarian Literature', in *Journal of Mediterranean Studies*, 5(2), 222–38.

Bernard, C. and Digard, J.-P. (1986), 'De Téhéran a Tehuantepec. L'ethnologie au crible des aires culturelles', in *Anthropologie: état des lieux.* Paris: L'Homme and Navarin, pp. 54–76.

Cattaruzza, M. (1993) 'Slovenes and Italians in Trieste, 1850–1914', in M. Engman ed., *Comparative Studies on Governments and Non-Dominant Ethnic Groups in Europe, 1850–1940.* Dartmouth: New York University Press, pp. 189–219.

Čapo Žmegač, J. (1999) 'Ethnology, Mediterranean Studies and Political Reticence in Croatia: from Mediterranean Constructs to Nation-Building', *Narodna umjetnost. Hrvatski časopis za etnologiju i folkloristiku*, 36 (1), 33–52.

De Franceschi, C. (1989) [1926], *Uspomene*, Pula and Rijeka: Čakavski sabor et alii.

Dragadze, T. (1987) 'Fieldwork at Home: the USSR', in A. Jackson ed., *Anthropology at Home*, London and New York: Tavistock, pp. 154–63.

Fikfak, J. and Johler, R. (eds) (1998), *Folk Culture Between State and Nation. Ethnographies at the Turn of the Century in Central Europe and 'The Austro-Hungarian Monarchy in Word and Image'.* Ljubljana and Vienna: Inštitut za slovensko narodopisje ZRC SAZU and Institut für Volkskunde der Universität Wien.

Geiss, I. (1994) 'Great Powers and Empires: Historical Mechanisms of Their Making and Breaking', in G. Lundestad (ed.), *The Fall of Great Powers: Peace, Stability, and Legitimacy.* Oslo and Oxford: Scandinavian University Press and Oxford University Press, pp. 23–43.

Grbić, J. (1997) 'Etnološki diskurs prošlosti – Kukuljević u Bosni 1858', *Bosna franciscana*, V(8), 145–55.

Grdina, I. (1999) *Od rodoljuba z dežele do meščana.* Ljubljana: Studia humanitatis.

Gross, M. (1973) *Povijest pravaške ideologije*, Zagreb: Izdavački servis Liber.

Hadžijahić, M. (1990) *Od tradicije do identiteta: Geneza nacionalnog pitanja bosanskih Muslimana*, Zagreb: Islamska zajednica Zagreb.

Hann, C. (1991) 'Europe centrale et orientale', in P. Bonte and M. Izard (eds), *Dictionnaire de l'ethnologie et de l'anthropologie.* Paris: Presses universitaires de France.

Hroch, M. (1985) *Social Preconditions of National Revival in Europe: A Comparative Analysis of the Social Composition of Patriotic Groups among the Smaller European Nations.* Cambridge: Cambridge University Press.

—— (1995) 'National Self-Determination from a Historical Perspective' in S. Perimal (ed.), *Notions of Nationalism*, Budapest: Central University Press, pp. 65–82.

Kósa, L. (2001) 'Hungarian Ethnographical Society' in M. Hoppál and E. Csonka-Takács eds, *Ethnology in Hungary: Institutional Background.* Budapest: European Folklore Institute, pp. 1–9.

Kržišnik-Bukić, V. (1996) *Bosanska identiteta med preteklostjo in prihodnostjo,* Ljubljana: Inštitut za narodnostna vprašanja.

Kukuljević Sakcinski, I. (1997) *Izabrana djela.* Zagreb: Matica hrvatska.

Leal, J. (1999) 'Saudade, la construction d'un symbole. "Caractère national" et identité nationale', *Ethnologie Française,* 29(2), 177–89.

Magris, C. (1963) *Il mito absburgico nella letteratura austriaca moderna.* Torino: Einaudi.

Mann, M. (1995) 'A Political Theory of Nationalism and its Excesses' in S. Perimal (ed.), *Notions of Nationalism.* Budapest: Central University Press, pp. 44–64.

Moraldo, S.M. (2002) 'Introduzione. Sul sentiero della multiculturalità': Karl Emil Franzos' in S.M. Moraldo (ed.), *Karl Emil Franzos: Racconti della Galizia e della Bucovina.* Rome: Salerno Editrice, pp. 7–26.

Radić, A. (1936) [1899] 'Izvještaj o putovanju po Bosni i Hercegovini' in *Sabrana djela,* 1, *Narod i narodoznanstvo.* Zagreb: Seljačka sloga, 143–76.

Radić, S. (1993 [1908]) *Živo hrvatsko pravo na Bosnu Hercegovinu.* Zagreb-Sisak-Ljubuški: Consilium.

Rizvić, M. (1989) *Između Vuka i Gaja,* Sarajevo: Oslobođenje.

—— (1994) *Panorama bošnjačke književnosti,* Sarajevo: Ljiljan.

Roshwald, A. (2001) *Ethnic Nationalism and the Fall of Empires: Central Europe, Russia and the Middle East, 1914–1923.* London: Routledge.

Stagl, J. (1998) 'The Kronprinzenwerk – Representing the Multi-National State' in B. Balla, and A. Sterbling eds, *Ethnicity, Nation, Culture: Central and East European Perspectives.* Hamburg: Krämer Verlag, pp. 17–30.

Starčević, A. (1992 [1898]) *Istočno pitanje,* Zagreb: Hrvatska hercegovačka zajednica 'Herceg Stjepan'.

Stocking, G.W. Jr. (ed.) (1984) 'Functionalism Historicized: Essays on British Social Anthropology', *History of Anthropology,* II, Madison: University of Wisconsin Press, 5–21.

Stoianovich, T. (1994) *Balkan Worlds: The First and Last Europe.* New York and London: M.E. Sharpe.

Strassoldo, R. (1991) 'Meje in sistemi: Sociološke misli o Srednji Evropi' in Vodopivec ed., *Srednja Evropa* Ljubljana: Mladinska knjiga.

Thiesse, A.-M. (1991) *Écrire la France: le mouvement littéraire régionaliste de langue française entre la Belle Epoque et la Libération.* Paris: Presses universitaires de France.

—— (1997) *Ils apprenaient la France. L'exaltation des régions dans le discours patriotique.* Paris: Éditions de la Maison des sciences de l'homme.

Vivante, A. (1997) [1912] *Irredentismo adriatico: contributo alla discussione sui rapporti austro-italiani.* Genova: Graphos.

Chapter 4

How Large are the Issues for Small Ethnographies? Bulgarian Ethnology Facing the New Europe

Galia Valtchinova

Among the many books that try to answer the question 'what is anthropology?' there is one that contains the formula: 'small places, [but] large issues' (Eriksen 1995). According to its author, a Norwegian anthropologist, 'smallness' may range from really small villages to modern nation states. Sharing the assumption that ethnology is to the heterogeneous European tradition what social/cultural anthropology is to the Anglo-American one,[1] I believe that this definition fits our concern with small ethnologies. Is 'smallness' related to place, to particular and carefully circumscribed ethnographies, to local knowledge (Geertz 1981)? Is it a question of 'siting culture' (Hastrup & Olwig 1997) or, rather, an issue of standing point, of a gaze – the Levi-Strauss's (1983) remote gaze, or an immediate one – which does or does not allow us to speak about the cultures/society we study? Is it something else, related neither to place nor gaze, but to the knowledge-and-power nexus which – after M. Foucault (1982) brought it to the attention of social scientists – is increasingly resorted to in response of questions about structures, authority and hierarchies within the discipline?

This chapter does not seek to answer these basic questions but rather explores a peculiar view of 'smallness' in ethnology and more precisely, some of the theoretical and practical implications of the 'standing point' when 'native' and non-native ethnologists engage in research in the small South-East European country of Bulgaria. The use of 'ethnography' in spite of 'ethnology' in the title is deliberately provocative; it alludes to the ambiguity of the term as it is used within a 'small' national tradition of ethnology. In what follows, I shall try to explain how a national ethnography and *etnografija* stand for the theoretical elaboration of a discipline like anthropology. To this end, I shall provide a historical overview of the two disciplines which, born as varieties of *Volkskunde*, are now struggling to dominate the field as well as 'ethnology' and 'anthropology' in Bulgaria. I will focus on the negotiation, during the second half of the twentieth century, of specifically 'ethnographic' and 'folkloristic' issues. In this way, I hope to show the switch between a narrower and a larger significance of 'ethnography' in the process of redefinition of the boundaries of ethnology that has recently started, as a strategy to bring a 'small' national ethnology

1 This is a simplification of a long debate, as well as of more complex and various realities: for a quick overview see Goddard, Llobera and Shore (1994).

closer to Western standards. Finally, I investigate, through a confrontation of various approaches to 'small' society, the kind of 'smallness' that characterizes the research field of a Nation-State-bound ethnology.

Between 'Ethnography' and Folklore Studies (1889–1940s)

In Bulgaria, 'ethnology' was twice born as *Volkskunde* within a century. In the nineteenth century, it was born under the label of *narodoznanie /narodouka*, a calque from *Volkskunde*. *Narodoznanie* was seen as collecting record and study 'material and moral products of the folk's spirituality', and closely related to the 'national revival'. The first steps in this direction were taken in the second quarter of the nineteenth century, through incentives coming from Russian and South-Slav erudite men that promoted projects of Slav emancipation and unity.[2] Their interest met with the support of enlightened Bulgarians willing to endorse *Bulgarian-ness* as an ethnic identity and a cultural asset. Interest in the Bulgarian language was especially welcome to counter the orientation of Bulgarian-born elite to the more prestigious Greek identity that was predominant at the time. By the 1850–60s, *narodouka* enjoyed high interest among the burgeoning intelligentsia, and many renowned writers and journalists engaged in audacious projects in collecting and disseminating knowledge of 'oral traditions'.[3] This interest was also fuelled by more clear-cut ideas of political emancipation from the Ottoman Empire that took the increasingly 'revolutionary' overtone of struggle for independence.

Two emblematic figures of the movement for national liberation, G. Rakovski (1821–1867) and L. Karavelov (1834–1879), are considered as founding fathers of Bulgarian ethnography.[4] However, the actual founding father of Bulgarian ethnology was the eminent philologist Ivan Shishmanov (1862–1928). In 1889, the then 26-year-old professor from the University of Sofia explained the 'importance and the tasks of our ethnography' in his large introduction to *Zbornik za Narodni Umotvorenija*, the first periodical that was launched on that occasion. In that paper, which became the manifesto of Bulgarian ethnology, Shishmanov talks indifferently of 'ethnography' and 'folklore studies'. In fact, the first generation of field workers and scholars made almost equal use of both terms, though in the long run their theoretical heritage, when available, was associated mainly with *Folkloristika* [Folklore studies]. An attentive

2 Among those who contributed the most towards the 'awakening' of Bulgarians' interest for their own culture and history are Jurij Venelin (1802–1839), a native of a then Hungarian-dominated Subcarpathian Ukraine who choose tsarist Russia (for his complex trajectory see Andrieu 2006), and Vuk Karažic (1787–1864), the 'father' of the modern Serbian language.

3 P.R. Slavejkov (1827–1895), an outstanding journalist and writer, collected thousands songs, proverbs and fairy tales, and published hundreds of them as in various journals and newspapers under the header of 'people's oral traditions'.

4 Rakovski elaborated the first Bulgarian index of folk customs and guidelines for their collection (1859); he also authored fantastic theories about the origins of the Bulgarian people (which he traced back to Vedic India). A more critical approach to collecting ethnography is due to Karavelov, who published (1861) the first collection of Bulgarian antiquities comprising customs and oral traditions. For these early developments cf. Benovska 2000: 311–13.

reading of the first ten volumes of *Zbornik za Narodni Umotvorenija* reveals that already by the turn of the nineteenth century, 'ethnography' was used to denote the 'invisible' (sometimes qualified as 'dark') and patient work of collecting materials – what would become later 'primary sources' (as opposed to university teaching and scientific production). On the eve of World War I, being an 'ethnographer' usually meant 'doing fieldwork', while academic distinction was associated with 'Folklore Studies'. The labels of ethnography and folklore have been used interchangeably until World War II. Whatever the label, the 'science of the folk' looked back towards the past, and sought to uncover an older and 'purer' state of popular culture and popular spirit. During this period, the difference between ethnography and folklore studies relied mainly on the distribution of tasks within a common project, in which the leading role fell on philology. Generally speaking, scholarly pursuits and academic career remained associated with 'folklore', even when 'collecting material' and 'doing scholarship' mixed up in the activity of a single person, as in the case of Mikhail Arnaudov (cf. Arnaudov 1971).

Collecting and 'Inventing' Data: Approaches to Fieldwork

During the early phase, the task of collecting artefacts of 'popular culture' was sometimes understood in terms of 'producing' them. Thus, Rakovski used to forge evidence when he could not find it on the spot, a style for which he was criticized even by his contemporaries. However, the most remarkable example of folkloristic mystification – the 'Veda Slovena' epic circle (published between 1874 and 1881) – was the result of an international collaboration between a Bulgarian teacher, Ivan Gologanov, and his Bosnian-Serb mentor, Stefan Verkovic (1821–1893), a fervent follower of Illyrianism and promoter of South-Slavic ideas. As a whole, Bulgarian scholarship has kept aside from such enterprises; and it has refrained from political manipulation of South-Slavic epics, common to most Balkan nationalisms.

It was Bulgarian folklorist Mikhail Arnaudov who proved the forgery behind Veda Slovena. After the publication of the 'manifesto', facts of brutal invention of 'field' data were no longer recorded. However, the manipulation of field data could occur on various levels of the collecting process, what could be observed in the case of D. Marinov (1846–1940). Without scholarly training, this parish priest in Northwestern Bulgaria – a duty that kept him close to the Volk for decades – remained a tireless fieldworker even when he settled down in the capital Sofia, and was appointed as Curator of the People's Museum in 1906. The publication of his enormous corpus of field data over twenty years (1891–1914) offers multiple insights into the ethnographer's methods and conduct in the field. His interest in theory limited to evolutionism; Marinov tried to 'find out' data about archaic forms and ways of life where they were lacking, and employed various procedures to making older field data (Vassileva 1998, Benovska 2000: 316–21). Reading between the lines, one feels perhaps that such reconstructions of the imagined original 'purity' of customs and mores were as much attempts at establishing distance. Given the lack of real social distance vis-à-vis people studied and the un-relatedness of 'the Other' to the early

Bulgarian ethnographic project, 'Time' was the only available resource to create the distance that was needed to produce ethnographic knowledge (Fabian 1983).

In the interwar period, fieldwork was given a larger place inside academia and became related to the institutionalization of 'field study' in the student's curriculum. Teaching folklore at the department of Slavic philology led to the constitution of important archives of students' works (the 'diploma memoirs'), produced at the outcome of one- to three-month period of field work.[5] These archives consisted mainly of village ethnographies: students were encouraged to study their own native places, especially villages in remote regions. A non-negligible part of these memoirs are studies of refugee population, i.e. of Bulgarians coming from Greece or Turkey. This is an echo of a much broader trend that is attested mainly in the national ethnologies of countries-'losers' from the WW1, which had to face a 'catastrophe' of the national project and to deal with forced displacement of populations in the aftermaths of the war.[6] Halfway between folk-life monographic work and community study, these writings show a growing awareness about real social contexts.

Compared to the 'sister' disciplines in other South-East European countries, Bulgarian folklore studies kept a rather low profile when faced to the challenge of the refugee populations: the students' graduation works and the few publications by renowned ethnographers were far from characterizing mainstream ethnology, like in Hungary or Greece. As in the neighbouring Romania (perhaps a victorious country), the reaction to disillusionment with the national idea and the loss of war took the shape of an 'interior gaze', a scrutiny of the collective self. All the major writers, literary critics and theorists, artists, but also historians of the interwar period conducted or participated in heated debates on 'Bulgarian psyche' and the 'national soul', mixing in their arguments folklore, glorious past, cultural heritage, popular culture, ethnic territory, blood and 'racial traits'.[7] The issues of language, philology, and folklore have provided the core arguments of the great debate on modernity.[8]

5 There are two separate archives, both named after Professor St. Romanski who was the head of the 'folklore studies' within this Department during 1920s–1947, and his wife and collaborator, Professor (since the 60s) Tz. Romanska. The former, containing ca. 250 unpublished 'memoirs', was little known till the late seventies, when it was made accessible and put to contribution by home ethnologists. Both ethnographers and folklorists of the 1980s saw these 'memoirs' as source of reliable data for 'customs and practices' of the Interwar period.

6 The best example of this trend is Hungary, where studying Transylvania (annexed to Romania) became the corner stone of post-war ethnography and folklore studies: cf. Kürti 1997. Another example is Greece, where the most powerful scientific institutions assumed both the study and the preservation of the traditions and ways of life of the Greeks from Asia Minor (Prevelakis 1992; cf. Herzfeld 1982).

7 For data and analyses see Daskalov & Elenkov (eds) (1994) and Kiossev (1995) who stress the general 'right' political orientation of these scholars and intellectuals.

8 Like elsewhere in Southeastern and Central Europe (Verdery 1991: 46–71), in Bulgaria folklore and the discourse of folklore have been appropriated by the anti-modernizing trends.

Towards a 'Marxist Ethnology'

In the aftermath of World War II, another ethnology was born under the label of *etnografija* [ethnography], an emulation of the Soviet discipline developed under the same name. Borrowing from Soviet *etnografija* did not mean, however, following its theoretical ambition and scope of fieldwork, as defined by Bromley (1980), whose *Ethnos and ethnography* (1972) had perhaps become the new Bible of Bulgarian ethnographers. An outsider would never be tempted to say that Bulgarian *etnografija* was a kind of anthropology, as Gellner (ed. 1980) did for the Soviet one.

The discipline deserved a new institutional framework. In 1947, the Institute of Ethnography was founded at the Bulgarian Academy of Sciences, to coexist with the much older People's Museum (1906). The first chair of *etnografija* opened at the Department of History of the University of Sofia only in 1972. Integrated in the fabric of historical sciences, ethnography had its place as a 'complementary historical discipline'. Indeed, the new Bulgarian ethnology adopted historical approach as its main method, privileging the search for the origins and the primitive state (of an object, a practice, or a ritual). Already characteristic of early ethnography and folklore studies, this approach, now cautioned by Soviet science, was associated with the art and methods of 'reading' artifacts as crystallizations of the past. The material object came to be seen as a direct 'testimony' from the immemorial time of origins, thus functioning in a similar way as 'historical records'. It also privileged the material (against the verbal or the spiritual) dimension of tradition. For decades, *etnografija* celebrated the material framework of 'popular life' and work – a vision well fitting the axiom of determinant character of material world and of work in the Marxist social theory. Turning to field work, this conception of ethnography reinforced the vision of peasant culture as reflecting a primitive, 'initial' state of things, an idea that was already present in the way Eastern and Central European intellectuals constructed the nation (cf. Verdery 1991: 54–63). This led to multiplication of minor studies on various items of traditional culture, viewed as 'remnants of immemorial past' or, rather, beyond historical time. The village remained the primary unit of ethnographic research despite the tremendous modernisation and the rural exodus that depopulated the countryside in the second half of the twentieth century.

Etnografija neglected concerns with structure or functions of culture but valued the study of rapid social change. The economic and social changes due to industrialization and rural depopulation made it urgent to study not only the peasant milieu and material environment, now seen as the depository of a 'dying' way of life, but also the ways people used to cope with 'traditions' to which they were allegedly attached. This emphasis on village-scale research was due to pragmatic concerns with 'siting' the nation: since 'culture' was conceived as the charter of national essence, the village-scale work was conceived of as putting 'bricks in the wall' of the monolithic Nation., Taken as an epistemological unit, the village fitted the national construct; it contributed to the knowledge of the national model and at the same time it served to confirm it.

Starting from the late 1960s, Bulgarian *etnografija* progressively freed itself from its obsession with material charters of tradition, and turned a 'spiritual' interest for rituals, customs, feasts and beliefs. In fact, the latter were not rejected but only underestimated so as to suit the communist ideology and political discourse. The renewal led to the recognition of the rich heritage of folklorists and to the 'rehabilitation' of the great figures of pre-war folklore studies, previously rejected on the ground of being part of a 'bourgeois' and 'idealist' science. The best example of this tendency is the 'rediscovery' of Arnaudov, and especially his work on rites and customs, which dominated *etnografija*'s shift to religious feasts and ritual practices in the early 1970s.

Despite emphasis on oral literature and more broadly, oral 'production' as a charter of the popular tradition, Bulgarian *etnografija* never considered the spoken word and language as issues central to its subject matter. This gap was filled by the foundation of the Institute for Folklore within the Academy of Sciences in 1974, and the launch of the journal *Bulgarian Folklore* a year later. Despite their long tradition in academia, Folklore Studies was now granted a place within the 'pure' scientific network that flourished under socialism. Recruited mainly from the milieu of philologists, the new structure reasserted the role of oral tradition and the uniqueness of language as a charter for [folk] culture, without forgetting other, non-scriptural media as music and 'plastic arts'. The attention given to the issue of language, as well as the rehabilitation of literature in the widest sense of the term (including oral literature) determined the choice of the leaders of the new structure.[9]

The foundation of the new Institute induced competition over field and topics of research: both structures had the primary task of 'doing fieldwork' and to fill archives with field materials. 'Ethnographers' and 'folklorists' were committed to doing fieldwork in equal degree and in a similar way by teams whose members used to cross-check field data, through field trips in one or several villages of an area, for a duration between one week and one month. However, institutional competition led to claims for different methodologies and different approaches to the 'subject matter' of the discipline. If 'people's culture' was defined as the prime focus of *etnografija*, a shift from 'folk culture' to 'the folk human being' [folkloren čovek] was operated within folklore studies. In the course of the 1970s, home made theories (based on Russian and Western theoretical background) tried to describe various ways of handling 'field' material; in this enterprise folklorists were more dynamic and more open to new field experience and to collaboration with foreign partners.[10] Advantaged by the tradition of field training at the Slavic philology department of Sofia University, within two decades folklorists gathered a wealth of oral data and of music records, which gave the Folklore Institute Archives a weight similar to the older Archives for Popular Knowledge. Staying closer to the guidelines of Soviet

9 Petâr Dinekov, a brilliant linguist and internationally known scholar in the field of Old Bulgarian literature, produced an overview of Bulgarian folklore (Dinekov 1972) and was appointed Head of the Institute from 1974 to 1986. The second director, Todor Iv. Zhivkov, distinguished himself as scholar of oral literature (Zhivkov 1977) before moving to theory of folklore and (Zhivkov 1987), and a final synthesis of the new directions of 'ethnology' with regard to the nation (Zhivkov 1994).

10 A good example in this respect is the collaboration with the French Musée des Arts et Traditions populaires (ATP) launched in the late 1970s; the result was, among others, the development of the concept of 'small town' as an appropriate research unity for cultural change.

ethnology without having the audacity of its theoretical constructs, 'ethnographers' kept looking backward. (Genčev 1984)

As a whole, the understanding of culture by ethnographers differed little from that of the folklorists. Culture was conceived either as a core characteristic of a deliberately ethicized people/nation, or as an array of 'arts' and techniques performed in order to transmit the stock of memory, knowledge, and practices to future generations. Both celebrated a national model that was deeply rooted in the past; within Europe these two trends exemplified the practice of a 'small', i.e. nation-bound, ethnology.[11] Such a dependency upon a national project deserves special attention. As M. Herzfeld (1987) has argued in relation to modern Greece, Folklore Studies are predicated on the past, and focuses on peasant culture where the national project wins acclaim from History. This is clearly the case with most Balkan ethnologies. On the other side, as K. Verdery (1991: 54–70) has argued with reference to Romania, the focus of 'the Nation' creates a field of competition between various disciplines claiming to represent the national values better. This kind of competition has brought many intellectuals to the fore while a 'scientific knowledge of ethnic reality' remained at the heart the struggle for symbolic capital both within and outside academia. The two trends met and fused, producing a peculiar competition over certain 'slices' and events of the past deemed 'national': the ones that contained most prestigious History. It was under socialism, in the 1970s, that most of such 'ethnographic knowledge' was produced.[12] Despite the claims of 'dialectical-materialistic' vision and the historicity which was inherent to it, the focus on Nation remains unchanged and the understanding of people's culture, an essentialist one. In this context, one should not be surprised with the development of semiotics in ethnology under socialism: in the seventies and the eighties, Bulgarian ethnographers and folklorists embraced the methods and the theories of representatives of the 'Tartu school', like Vs.V. Ivanov, V.S. Toporov and F. Uspenskij. The semiotic approach allowed them to seek explanation of cultural facts recorded *hic et hunc* in going away from present-day social contexts, all the more by keeping in line at least formally with 'Soviet' theorists.[13] With the reassessment of the role of social sciences, following the 1989 political change, this would-be innovative methodology revealed itself to be quite an uncomfortable starting point for research.

Bulgarian Ethnology after 1989

Since 1990, Bulgarian ethnology has been subject to a tremendous change which is evident in all fields, structures and institutions, and at all levels of ethnological

11 The structural resemblances between nation-bound ethnologies (the German case compared to the French one) are well delineated by comparison to a colonial-bound ethnology in France: cf. Brückner (1987). Taking for example the Greek laographia, M. Herzfeld has explored the issue in two books (1982, 1987), within a still larger framework.

12 In Bulgaria, it was mainly a search for survivals of Thracian antiquity in religion (Valtchinova 1998).

13 Essentialism is inherent to the theory of the ethnos of Yu. Bromley, the highest authority in Soviet etnografija (Bromley 1972, 1980): cf. Gellner (1988).

practice and theory. The change of political regime that began in November 1989 led to a radical change of paradigm, a change concomitant of political re-orientation that developed in several directions. While many of the issues studied during the socialist period turned to be obsolete, new issues and methods emerged and quickly imposed themselves. This change was manifested in the re-discovery of the familiar 'ethnoscape' (as defined by Appadurai 1996: 48) and the discovery of new social landscapes, in 'opening the eyes' for the immense cultural diversity that challenged the nation-bound ethnography. It led, progressively, to a decisive turn towards social diversity and the tremendous social change, as well as the study of new social forces and specific actors. This change affected both the internal structure and the length of the research field of Bulgarian ethnography. The introduction of new strategies in the institutional politics also led to a large-scale revision of scholarship.

The Change of Paradigm: Fieldwork, Text, and Themes

One of the most significant changes in the paradigm of ethnological study is the care to circumscribe one's own research in present-day realities, and to take the latter into account when trying to explain the processes studied. This leads to a neat retrieval from reference to remote or 'immemorial' past as explanatory device, and to increasing attention toward short-term historical change. New concepts of doing fieldwork imported from various Western anthropological traditions have brought about a critical rethinking of previous field experiences and of ways of organizing and conducting fieldwork. Here and there, concerns have been voiced of whether the ethnographer's presence on the field must be noted through a series of formal criteria such as cross-reference with materials gathered by other ethnologists on the same field site; by specifying the number of informants interviewed; and above all by giving the names and personal data of informants and refusing to keep them anonymous. Compared with the long-term field work on a single place – a *sine qua non* of Western anthropological training – the requirements of the 'stationary work' of Bulgarian ethnographers seemed ridiculous, or at best outdated. The obvious differences between the native scholar's problematic and field methods, and those of Western-born or Western-trained anthropologists doing fieldwork in Bulgaria, have challenged the native ethnographer's certainties about his/her exclusive position in producing knowledge of his/her own culture. Ethnographers helped raise questions about distance, insider-hood, rooting in a community, of the remote gaze, etc. – questions that, even not ignored, were rarely addressed.

If 'anthropology' is defined through an implicit or explicit commitment to the methodology of ethnographic fieldwork, which is seen as generating a unique and valuable form of knowledge (Wolfe 2000: 197), subscribing to 'anthropology' in post-socialist Bulgaria indicates a tendency towards generalizing and 'theory-making'. The flow of new and exciting theories that became known in relation to renowned anthropologists and ethnologists led 'newcomers' to the discipline to underscore the importance of fieldwork. Thus, a twofold process took place (and it is continuing in many respects): on the one hand, people who trained as sociologists, literary critics, or cultural scientists were attracted by the theoretical sophistication of British and American-style anthropology, which they viewed as fitting to their own expertise. The post-modern trend and the debates about 'writing anthropology' led some of these experts to exercise 'anthropology' on texts and discourses, though they had never set foot on any kind of 'field'.[14] Likewise, most of the experienced ethnographers who had matured under socialism were not inclined to theorize, and were slow to adapt to the new theoretical diversity. This discrepancy (as well as the entirely new requirements for financing projects and programs) led to the constitution of multi-disciplinary teams composed of historians, sociologists, political scientists; even if based on fieldwork, the theoretical output resulting from this research was rarely written by the fieldworkers.[15] By 2000, however, the necessity for fieldwork (conducted in a variety of ways, with varying duration) was acknowledged by those engaged in research or teaching of 'ethnology' or 'anthropology'

The change of paradigm became visible in the emergence of new issues. Studying 'culture' (intended here as 'things', techniques, and artifacts) is no longer at hand. Although books and articles in line with the old paradigm were still being published, new domains of ethnological research have been developing since the early 1990s. The issues of transition, democratic change or ethnic/cultural diversity, captured the attention of Bulgarian ethnographers and former students of folklore. Promoted by international agencies and providing funding opportunities without common measure to local research, they focalized the significant encouragement coming from outside. Almost absent during the previous period, research on social change became one of the most exciting trends. It was oriented towards new social forces and new political mentalities, as well as to specific actors of the transition period such as entrepreneurs, smugglers and the market (Aleksandrov 2001; Konstantinov 1996; Konstantinov, Kressel, Thuen 1998); 'fighters' and 'bodyguards' (Ivanova 1999). This new research seeks also to shape a more general idea of a society in transition, starting from the ethnographic criteria of kinship, social structure, solidarity, clientelism and political parties (cf. Benovska 2001).

Another facet of the change concerns the scope of analytical work. Most of the native ethnologists went from studying nation-wide cultural practices to community study. This shift was rather significant since nation-wide study implied a command of materials gathered across the whole of the 'Bulgarian ethnic territory' and within a century at least. 'Community' or regional study was seen as a means for delineating idiosyncrasies within the otherwise uncontested national model, thus contributing to its deeper knowledge. Since 1989, things have dramatically changed. The nation-wide study of cultural characteristics has lost both its attraction and its normative strength, while case-studies and community-centered study have been widely practiced. Increasingly, 'home' communities have become particular arenas that exemplify past and ongoing processes of economic and political change (cf.

---

14 A good example in this sense is Kiossev (2002) on a concept inspired by Herzfeld (1997).

15 This tendency is avoided by the largest socio-ethnographic study based on field work, which was realized with EU support in the first half of the nineties: Zhelyazkova ed. (1995). Yet in most collective works sociologists are the preferred partners, for their ability to bridge between fieldwork and theory: cf. also Zhivkova ed. (1996), Giordano, Dobreva and Lohman-Minka eds (2000).

Wolfe 2000: 202–10), rather than simply arenas of still living cultural practices rooted in the past. Should one compare this process with the post-war emergence of community studies in American sociology and anthropology, dominated by the figure of R. Redfield, they would realize that both enterprises obey to different logics. In the new Bulgarian ethnology the shift to community studies has had its starting point in the nation: community is approached 'from above', while Redfield and his followers did it 'from below'. This complicates the change of a nation-centered ethnography: given past developments, such an approach leaves the door open for milder and veiled forms of celebration of the national project through national community. Since community-centered study has to be well circumscribed in the national space and vis-à-vis main loci of power, there is a real risk for the 'native' ethnologist working in the capital city to remain blind for his own role as producer of authoritative discourse. Such a risk is inherent to the framework of producing academic knowledge by a 'small' ethnology: working within national territory, with a focus on parts and portions of the nation, is still the dominant way of envisaging ethnography. Bulgarian ethnologists could overcome the trap of a national–bound ethnography by going outside the 'home society' and 'home' State, and by engaging in various fields in and outside Europe – or by searching the Other nearby. In order to give an idea of the logic of these new orientations, let us now turn to the issue of ethnic diversity and minorities.

Discovering Diversity: Turks, Gypsies, Pomaks and Others

The study of ethnic and ethno-religious groups and communities emerged powerfully in the aftermath of political change and within a few years, it became a major issue. Before 1989/90, cultural diversity received little attention, if any, from either ethnographers or folklorists. There seemed to be no place for multiethnic and multicultural vision in the tasks of a national ethnography and a national model of doing fieldwork. Other ethnic and/or religious groups were given attention as far as studying them enabled the shaping of a certain theory of the ethnos and 'ethno-social organism' following Soviet *etnografija*. Thus viewed, a study of them differed little from the study of 'ethnographic groups'.

Following the theory embraced during socialism, the 'ethnographic groups' were characterized by a range of specific 'cultural traits' which made for their difference within a national 'model' and created their specificity (perceived by both outsiders and insiders), without challenging the group members' collective consciousness as being part of the Bulgarian nation. As a rule, ethnographic groups share the Christian-Orthodox confession characteristic of the Bulgarians, the only exception being Muslim Pomaks.

Ethnic groups were another sub-category in this construct, ranging just below the basic category of 'nation'. Contrary to the 'ethnographic group', the members of an ethnic group 'possess a separate collective (ethnic) consciousness'. The distinction made in terms of 'national consciousness' is an important one: 'being conscious' about one's own group difference from the dominant nation could challenge the 'national model'. In this sense, past research on these cultures has sought to smooth away differences and to underline similarities with 'the Bulgarian culture'. Study of ethnic groups concentrated upon formal characteristics, and to the features, which, especially with regard to Muslim cultures, managed to veil their essential Otherness. Following this vision, attention to ethnic groups was initially limited to the collection of songs, tales and other forms of 'expressive language'; later, it consisted in recording rites, customs and beliefs relative to various domains of 'popular culture' implicitly defined against 'the Bulgarian national norm'. This was only during the 'national regeneration' process forcibly imposed to the Bulgarian Turks, in the eighties, that Bulgarian ethnographers and folklorists turned to a more thorough study of the Muslim others within the nation. It must be said that the role of ethnologists and folklorists for launching this action and in the ongoing process remains ambiguous. No doubt, some of these studies comforted the vision of the 'backward Muslims' who need to be 'enculturated' and 'civilized' through their adhesion to the modernizing (i.e. explicitly Bulgarian) model. Local ethnographies promoted also the vision of the importance of pre-Islamic strata in the culture and everyday life of Turkish and Pomak people, thus contributing to the political thesis of forced Islamicization and Turcification of the 'Bulgarian Turks', which constituted the cornerstone of the 'national regeneration' process.[16] Within a few years, everyday life, women's life, Islam and its importance to shape the worldview of this population emerged as issues to be studied.

Since 1989, speaking of ethnos has become obsolete, and the whole construction of ethnicity has undergone a radical change. The emergence of a new paradigm in the field of ethnicity and cultural diversity affected above all the so-called ethnographic groups, leading to the suppression of the category itself. This intermediary level between an ideally homogenous Bulgarian whole, and ethnic groups enriched by some kind of essential Otherness, had no more place in the vision of ethnic and religious communities which coexisted for centuries before a national idea was superimposed to it: a kind of multiculturalism *avant la lettre* (Roth 1999). This radical shift from an essentially nationalist to a multiculturalist vision became known as the *komşuluk* thesis, or the theory of the peaceful coexistence of ethnic and religious groups, living in neighbourhood on a daily basis. Going back to the Ottoman times and realities, the term of *komşuluk* was especially useful in stressing the basic non-rejection of the Muslim and the 'compatibility' between Muslim and Christian populations (Zhelyazkova (ed.), 1995). It became a new codeword for the sociological and ethnological research of the nineties, the cornerstone of the political concept of a specific 'Bulgarian ethnic model' of peaceful transition to democracy. It should be noted that the *komşuluk* (and the realities behind it) was singled out by social scientists who studied the implosion of Bosnian multicultural society during the war (Bougarel 1996: 99–125). Increasingly challenged from the outside (Roth 2006), the *komşuluk* thesis is now further elaborated in academic teaching.

16 The 'national regeneration' process of 1984–89 was the major violation of human rights in socialist Bulgaria. Since its last phase coincided with, and in a sense accelerated the political change of 1989, it became one of the most researched topics: see for example Alhaug & Konstantinov (1995), Fotev (1998), Poulton (1991), Zhelyazkova ed. (1995), Zhivkov (1994).

Turning back to ethnographic groups of the old classification, one may say that the whole classificatory system articulating the internal variety of the 'Bulgarian ethnos' was no more suitable to the new research perspective. Some of the ethnographic groups did simply disappear from the vision of ethnologists; others were given the status of separate ethnic groups. The second move reflects both ongoing processes of rethinking the notion of ethnicity, and the very dynamism of democratic change and new political conjuncture. The case of the Pomaks is illuminating in this respect: immediately after the democratic change, ethnologists, sociologists and political scientists started to record memories of the recent events and to study their new sense of identity. As a rule, these studies focus on Pomaks' difference, as well as diverging processes vis-à-vis the 'Bulgarians', and recast them against other Muslim groups.[17] The ethnic group of the Gypsies (or Tsiganes) has also become an attracting research topic for ethnographers, sociologists, and political scientists. Their attention has rapidly materialized in a range of books and collections of papers, which contrast with the lack of ethnographical and ethnological writings of the previous period. The new vision of ethnic groups reaches its height in the panoramic view of all communities living in Bulgaria provided by Krasteva (ed., 1998a) whose study marks a decisive shift from the previous understanding of ethnic groups and the essentialism inherent in the nation.

The Sites of Change

The new developments delineated above brought to the redistribution of 'capital' (in the sense of Bourdieu's cultural or symbolic capital) between the sites of research. First, the importance of field sites had changed dramatically, both for their inhabitants (and potential or real 'informants') and for the researchers trained for fieldwork in 'other places'. If, until recent, interest for 'traditional Bulgarian culture' and 'the folk man' had brought teams of folklorists and ethnographers in remote villages where the only challenge was to speak the local version of Bulgarian correctly, today the substantially renewed teams have to assume fieldwork among Pomaks, in Turkish villages and Gypsy encampments, resorting to the vernacular language of the group.

Speaking of the sites of fieldwork, there is a noticeable shift in the location from 'mono-cultural' Bulgarian villages to culturally mixed villages; for more than fifteen years now rural Muslim population, especially the Pomak and Turkish villages in the Rhodopes Mountaines, enjoy sustained scholarly attention. Small to middle-sized towns with high historical capital like Bansko and Melnik (Southwest Bulgaria), Assenovgrad (formerly a centre of Greek culture and identity, now it contains a large proportion of Muslims); Chiprovtsi (an old centre of Bulgarian Catholics), and Kârdzhali (with an overwhelming Turkish population) are other examples of attractive sites. Research in larger urban contexts (cities like Plovdiv or Varna), which display a highly mixed population, also enjoys an increasing popularity.

17 On the Pomaks, see Balikci (1998), Konstantinov (1997) (both outsiders to Bulgarian ethnology stricto sensu), Fotev (1998) (a sociologist), Georgieva in Krasteva (ed.) (1998) and Zhelyazkova (ed.) (1995).

The institutional change presents a picture of struggles and complex strategies for survival: here, one can speak of a true redistribution of power. Already at the beginning of the nineties, the label of 'anthropology' became very demanded in scientific quarters, while experts with various education and training, who became public personalities and media figures, loved to call themselves 'anthropologists'. Impulses for change came from both inside and outside academic ethnology – from sociologists, political scientists, historians of literature and literary critics, and 'pure' historians. The old institutional sites of ethnological research remained, not without significant efforts to keep their authority.

Among the structures within the Bulgarian academy of sciences, the Institute of Folklore has been more adaptive to the new trends. It could be partly due to more intense exchange with Western scholars, especially in the framework of long-term research programs focused on small towns (Bansko, Chiprovtsi), attractive places of change.[18] Its journal (*Bâlgarski Folklor*) took the subtitle of 'Journal for folklore, ethnological and anthropological research'; it encouraged research on innovative topics, and produced special issues such as 'The paranormal' (1993), or 'Biography and life history method' (1995), which became the focus of interest from other social sciences as well as important arenas of theoretical exchange.

Though less visibly, the Ethnographic Institute and Museum too underwent a significant restructuring process of their structure and production. The name of the Institute's journal changed from 'Bulgarian ethnography' to 'Bulgarian ethnology', yet the series of regional studies it kept publishing is still marked by ancient trends of studying 'popular culture'. Apparently less touched by 'anthropology' fashion, the books and periodicals published by the Institute of Ethnography show a growing concern with theory and a new understanding of ethnology (cf. Tsaneva 2000).

Universities are other institutional loci of change where students' interest for ethnology, and especially for 'anthropology', makes the task more urgent and solutions more numerous. Here, the change of paradigm took less time (the pressure from young people being a powerful stimulus in this sense), and it was experienced in three ways: by creating new formal structures that corresponded to new content; by adapting new content to ancient forms; and by adopting a new institutional label at the price of minimal change in the content of an already existing course. The most interesting examples come from the University of Sofia and the New Bulgarian University, also in Sofia. As regards the former, the Department of History named the chair of Ethnography 'Institute of Ethnology' which was created within the Department of Theory and History of Culture (Faculty of Philosophy) in the 1980s. This nomenclature and the competence of the teaching staff (historians and philosophers) were disputed by the leader of Folklore studies at the Slavic Philology Department: by 1996, the University Council had to decide which of the two 'cultural anthropologies' was to be validated. It is difficult to say whether the final decision (favouring the former centre) was due to a clear vision of what 'anthropology' is, or to personal networking and to political capital of the personalities involved in the contest.

18 Relying on this collaborative work, French scholars have produced significant opuses: cf. Cuisenier (1998) (the Bansko programme), Chevalier (2001) (the Chiprovtsi programme).

Created and run with the support of the Open Society Fund, the New Bulgarian University is a place of experimentation *par excellance* where several formulae of 'anthropology' have been launched since 1992. The discipline has been refracted through literary studies and literary criticism, a domain in which the commitment to cultural studies had its earliest manifestation (Al. Kiossev). Another way forward for 'anthropology' is 'political anthropology', a combination of political science and sociology (V. Garnizov, E. Dajnov), and its participation in several well-funded programs that promoted alternative visions of ethnographic field (i.e. "doing fieldwork on NGOs"). Another path was 'religious anthropology' (taught by philosophers), a highly speculative subject which had little to do with social sciences or religion. The 'basic programme for Anthropology' launched in 1994–95 brought together former and current members of the Institute of Folklore, the Department of Classics, and other institutions engaged in cultural studies, mostly philologists and philosophers. A truly anthropological path, in the sense of American anthropology, is followed by the 'Institute of Anthropological Field Research' created by Yulian Konstantinov in 1997 in the framework of the NBU. The resulting studies on minority groups in Bulgaria, on tourist-traders (Konstantinov 1996, 1996; Konstantinov, Kressel & Thuen, 1998), on Russian reindeer herding (Konstantinov 2004) use to be funded by international agencies and enjoy international interest. In spite of the innovative field methods (cf. Herzfeld 2001: 152–54), this research had little impact on the developments of the discipline at the University.

Finally, research foundations and newly founded granting institutions (mostly US-sponsored) have deeply influenced scholarly orientations. The International Centre for Minority Studies and Intercultural Relations (IMIR) deserves special attention: through funding fieldwork in Bulgaria and elsewhere in the Balkans, and also by publishing the results of the sponsored research promptly, it contributed to the change. By its activities and the concept of 'urgent anthropology' (Zhelyazkova 2001), the Centre became increasingly visible in academic arena.

The Changing Source of Ethnographic Authority: the Self vs. the Other

In a renowned article on Greek 'laographia', the geographer Prevelakis (1992: 75–6) explains that though "studied by the others", Greece tends to refrain from studying foreign peoples because "she doesn't have colonialist ambitions," and because financial efforts to support research in remote countries is too big a burden for small states. Both assessments hold true for other 'small' national ethnologies of South-East Europe. Yet, however incontestable, the argument of lack of financial support for true anthropological work veils the opposition between studying oneself and studying 'others'. It is not enough to blame ideology implicit in *laographia*, *etnografija*, or another hypostasis of Folklore Studies; refusing the nationalist posture and point of view should be followed by practice outside, in places irrelevant to the 'Nation'. What brings us back to the intellectual genealogies of small Balkan national ethnologies – all of them varieties of *Volkskunde* – in national projects, and not in imperial ones paired with colonialism, as suggests Prevelakis again in his former argument.

Since its beginnings, Bulgarian ethnography/folklore studies has focused on the 'own' culture. In this viewpoint, it is 'self-ness' that matters, not otherness. This positioning vis-à-vis the actors of society (called the 'holders' or 'carriers of culture') under scrutiny structures the field and shapes the subject of the discipline in Southeast Europe in quite a different manner from, say, French ethnology or American cultural anthropology. The distance that subsisted, a social one, has never been central to the mental work and organizational framework of the discipline as it had developed in Bulgaria. This perspective is the exact opposite of the (cultural or social) anthropological project in the strict sense of the word, which initially excluded the object/subject relationship within the same culture or society. More, it did not allow space to the 'remote gaze' that Levi-Strauss placed in the heart of the anthropological project. The remote gaze of the others – those who do not share the socio-cultural and mental background with the culture studied, just try analyzing it – tended to be disqualified: the more remote the gaze, the less accessible the 'deep knowledge'. As Herzfeld repeatedly observed in his *Cultural Intimacy* (1997), even sharing a certain degree of cultural intimacy does not make the Western anthropologist 'ours'. The very posture of understanding and analytically explaining 'a culture' is in stark opposition to the national ego-centered perspective, and the latter requires as much 'understanding' as celebration of one's own culture.

As a native Bulgarian ethnologist, I see two ways for neutralizing the siren song of *Volkskunde*, and pursuing a truly ethnological project. One is to search for the Other, for cultural and religious otherness, within one's country or at its margins. The national self is deconstructed in a way that allows cultural diversity to be conceived of in its dynamics, and appear behind a monolithic and a-temporal identity. The other is to search for contexts of fieldwork in which the ethnographer can be a 'foreigner' and experience outsider-ness, multiplying techniques of estrangement – i.e. that side of the ethnological practice for which 'homecoming' (Kürti 1996) is not represented enough. This means not simply exploring remote countries but also turning to the boundaries of one's own country, and produce ethnographies of the making and remaking of national borders. Thus, the "national experience" is looked at from the 'other' side', and is tracked down in unexpected places where "small differences become large issues, and 'Selves become Selves through experiencing Others' (Linde-Laursen 1997: 160).

Current developments suggest that *etnografija* and Folklore Studies are likely to merge together gradually and form a 'small' European ethnology, switching from a national to a European perspective. Drawing on the whole array of social sciences and humanities, the two competing disciplines can only meet and fuse in a common project that is less 'national' in its field and more universal in approaches and issues.

Bibliography

Aleksandrov, H. (2001) 'Le prix du succès: être entrepreneur en Bulgarie aujourd'hui', in *Ethnologie Française*, XXXI (2), 317–27 [special issue 'Bulgarie']

Alhaug, G. and Konstantinov, Y. (1995) *Names, Ethnicity and Politics: Islamic Names in Bulgaria 1912–1992.* Oslo: Novus Press.

Andrieu, Chr. (2006) 'Jurij Venelin (1802–1839): les ambitions du découvreur de la langue bulgare en Russie', *Slavica Occitania*, 22, 45–61.

Appadurai, A. (1996) *Modernity at Large. Cultural Dimensions of Globalization*. Minneapolis and London: University of Minnesota Press.

Arnaudov (1971) *Studii po balgarski obredi i legendi, Sofia. Bulgarski Pisatel* [reprint by 'Prof. Marin Drinov' Academic Publisher, 1999].

Balikci, A. (1998) 'Symbolic geography of the Pomak', in G. de Rivas (ss. la dir.) *Nationalités et minorités en Europe: Forum Europe*. Paris : Eds. Tassili, 158–65.

Baskar, B. (1999) Les anthropologies face à l'effondrement de la Yougoslavie', *Diogène*, 188 (Octobre–Decembre): 68–84.

Benovska, M. (2000) 'The ethnology, an image of the world' [in Bulgarian], in *MIF (Mitologija. Izkustrvo. Folklor), 5: The Cultural Sspace*. Sofia: The New Bulgarian University, Dept. of History of Culture, 308–32.

Benovska–Sabkova, M. (2001) *Political Transition and Everyday Culture* [in Bulgarian]. Sofia: Academical Edition 'Prof. M. Drinov'.

Bougarel, X. (1996) *Bosnie, anatomie d'un conflit*, Paris: La Découverte, Etat du monde.

Boyadzhieva, St. (2001) 'Folklore, ethnographie, ethnologie: recherche et théorie en Bulgarie au XXeme siècle', in *Ethnologie Française*, XXXI(2) [Special issue: Bulgarie], 209–18.

Bromley, Yu. (1972) *Etnnos and Ethnography* [in Bulgarian]. Sofia: Nauka I Izkustvo.

—— (1980) 'The object and subject-matter of ethnography', in E. Gellner (ed.), *Soviet and Western Anthropology*, with an introduction by Meyer Fortes. London: Duckworth, pp. 151–60.

Brückner, W. (1987) 'Histoire de la Volkskunde. Tentative d'une approche à l'usage des Français', in I. Chiva et U. Jeggle (eds), *Ethnologies en miroir. La France et les pays de langue allemande*. Paris: Ed. de la MSH, pp. 223–47.

Chevalier, S. (2000) 'Stratégies d'échanges en Bulgarie', *Balkanologie*, IV(2), 59–71.

Cuisenier, J. (1998) 'Les Noces de Marko. Le mythe et le rite en pays bulgare'. Paris: Presses Universitaires de France.

Daskalov, R. and Elenkov Iv. (eds) (1994) *Zashto sme takiva?* [Why Are We Like That?]. Sofia: Narodna Prosveta.

Dinekov, P. (1972) *Balgarski folklor*, Sofia: Bulgarski Pisatel

Eriksen, T.H. (1995) *Small Places – Large Issues: An Introduction to Social and Cultural Anthropology*. London: Pluto Press.

Fabian, J. (1983) *Time and the Other. How Anthropology Makes its Object*. New York: Columbia University Press.

Fotev, G. (1997) *Drugijat etnos* [The other ethnos]. Sofia: The University of Sofia, St. Kl. Okhridski' Publisher

Geertz, C. (1983) *Local Knowledge: Further Essays in Interpretive Anthropology*. New York: Basic Books.

Gellner E. (1988) *State and Society in Soviet Thought*. London: Basil Blackwell.

—— (ed.) (1980) *Soviet and Western Anthropology*, with an introduction by Meyer Fortes. London: Duckworth.

Genčev, St. (1984) *Narodna kultura i etnografija*, Sofia: Nauka i Izkustvo.

Giordano, Ch., Kostova, D. and Lohman-Minka, E. (eds) (2000) *Bulgaria: Social and Cultural Landscapes.* Fribourg: University Press [Studia ethnographica Friburgensia, 24].

Goddard, V., Llobera, J. and Shore, C. (1994) 'Introduction', in V. Goddard, J. Llobera, and C. Shore, C. (eds), *The Anthropology of Europe*. Oxford: Berg. Harris, pp. 1–40.

Hastrup, K. and Olwig, K.F. (eds) (1997) *Siting Culture: The Shifting Anthropological Object.* London: Routledge.

Herzfeld, M. (1982) *Ours Once More: Folklore, Ideology, and the Making of Modern Greece.* Austin: University of Texas Press.

—— (1987) *Anthropology through the Looking-Glass: Critical Ethnography in the Margins of Europe*. Cambridge University Press.

—— (1997) *Cultural Intimacy: Social Poetics of the National State*. New York and London: Routledge.

—— (2000) *Anthropology: Theoretical Practice in Culture and Society*. Oxford: Blackwell.

Ivanova, R. (1999) *Folklore of the Change. Folk Culture in Post-Socialist Bulgaria.* CXXXIII2 (270). Helsinki: Academia Scientiarum Fennica.

—— (1998) 'New Orientations in Bulgarian Ethnology and Folkloristics', *Ethnologia Balkanica*, 2, 225–31.

Kiossev, Al. (1995) 'The Debate about the Problematic Bulgarian: A View on the Pluralism of the National Ideologies in Bulgaria in the Interwar Period', in K. Verdery, and I. Banac (eds), *National Character and National Ideology in Interwar Eastern Europe*, New Haven: Yale UP, pp. 195–217.

—— (2002) 'Dark Intimacy: Maps, Identities, acts of identification', in D. Bjelic and O. Savic (eds), *Balkans as Metaphor. Between Globalization and Fragmentation.* Cambridge: The MIT Press, pp. 165–90.

Konstantinov Y. (1996) 'Patterns of reinterpretation: trader tourism in the Balkans (Bulgaria) as a picaresque metaphorical enactment of post-totalitarianism', *American Ethnologist*, 23(4), 762–82.

—— (1997) 'Strategies for sustaining a vulnerable identity: the case of the Bulgarian Pomaks', in H. Poulton, S. Taji-Farouki (eds), *Muslim Identity and the Balkan State*, New York: New York University Press, pp. 33–53.

—— (2004) 'Towards a Model of Comparing Transitional Forms in Russian Reindeer Herding', Working Paper no.70. Max Planck Institute for Social Anthropology, Halle/Saale.

Konstantinov, Y., Kressel G. and Thuen Tr. (1998) 'Outclassed by Former Outcasts: Petty Trading in Varna', *American Ethnologist*, 4, 729–45.

Krasteva, A. (ed.) (1998a) *Communities and Identities in Bulgaria.* Ravenna: Longo Editore [Europe and the Balkans International Network, 10].

Kürti, L. (1996) 'Homecoming. Affairs of Anthropologists of and in Eastern Europe', *Anthropology Today*, 12(3), 11–15.

—— (2001) *The Remote Borderland: Transylvania in Hungarian Imagination.* New York, State University of New York Press.

Linde-Laursen, A. (1997) 'Small Differences – Large Issues: the making and Remaking of a National Border' in: V.Y. Mudimbe (ed.), *Nations, Identities, Cultures*. Durham-London: Duke University Press, pp. 143–64.

Poulton, H. (1991) *The Balkans. Minorities and States in Conflict*. London: Minority Rights Publication.

Prevelakis, G. (1992) 'La "laographie" grecque – ethnogéographie ou idéologie?', *Géographie et Cultures*, 2 (été), 75–84.

Redfield, R. (1953) 'The Little Community', in *The Little Community & The Peasant Society and Culture*. Chicago: The University of Chicago Press, pp. 1–48.

Roth, K. (1999) 'Toward Politics of Interethnic Coexistence'. Can Europe Learn from the Multiethnic Empires?, *Ethnologia Europaea*. 29(2), 37–51.

—— (2006) 'Living Together or Living Side by Side? Interethnic Coexistence in Multiethnic Societies' in R. Byron and U. Kockel (eds), *Negotiating Culture. Moving, Mixing and Memory in Contemporary Europe*. Berlin, Münster: Lit Verlag, pp. 18–32.

Shishmanov, I. (1889) 'Znachenieto i zadachite na nashata etnografija' [Importance and tasks of our ethnography], in *SbNU* [Zbornik Narodni Umotvorenija] I, 1–64.

Tsaneva, E. (2000) *Interpreting Ethnicity: Historiographical Overview and Assessment of Theoretical Discussions*. Sofia: Ethnographic Institute & Museum/ 'Assi'.

Valtchinova, G. (1998) 'Ethnographie et folklore du religieux en Bulgarie: un "tango de Lénine"?', *Ethnologia Balkanica*, 2, 145–65.

—— (1998) 'Anthropology of the Mediterranean and the Perspectives of National Ethnography' in A. Krasteva (ed.), *Communities and Identities*. Sofia: Petekson, pp. 186–203.

Vasileva, M. (1998) 'Dimitar Marinov. His Life and Work (October 14, 1846 – January 10, 1940)', *Ethnologia Bulgarica*, 1 [Sofia], 117–25.

Verdery, K. (1991) *National Ideology under Socialism: Identity and Cultural Politics in Ceauşescu's Romania*. Berkeley: The University of California Press.

Vermeulen, H. and Roldan, A.A. (eds) (1995) *Fieldwork and Footnotes: Studies in the History of European Anthropology*. London: Routledge

Wolfe, Th. C. (2000) 'Cultures and Communities in the Anthropology of Eastern Europe and the Former Soviet Union', *Annual Review of Anthropology*, 29, 195–218.

Zhelyazkova, A. (ed.) (1995) *Relations of Compatibility and Incompatibility between Christians and Muslims in Bulgaria*. Sofia: International Centre for Minority Studies and Intercultural Relations' Foundation (IMIR).

—— (ed.) (1998) *The Fate of Muslim Communities in the Balkans: Between Adaptation and Nostalgia*. Sofia: IMIR.

—— (ed.) (2000) *Albania and Albanian Identities*. Sofia: IMIR.

—— (2001) *Urgent Anthropology 1: The Albanian National Problem and the Balkans, Fieldwork*. Sofia: IMIR.

Zhivkov, T. Iv. (1977) *Narod i pesen* [Nation and Song]. Sofia: Bulgarian Academy of Sciences Publisher.

—— (1987) *Etnokulturno edinstvo i folklor* [Ethnocultural unity and folklore]. Sofia: Nauka Izkustvo.

—— (1994) *Etnicheskijat sindrom* [The ethnic syndrom]. Sofia: Alja.
Zhivkova, V. (ed.) (1996) *Malkijat svjat na socialnite procesi* [The Small World of social processes]. Sofia: Institute of Sociology.

Chapter 5

Challenges to the Discipline: Lithuanian Ethnology between Scholarship and Identity Politics

Vytis Ciubrinskas

It is risky to discuss ethnology as a scholarly discipline in Lithuania as it is often not considered as such *stricto sensu*. Ethnology is in fact referred to as a 'bogus science' sometimes, not only by lay people but also by historians. As the Lithuanian semiotician, and pupil of Claude Lévi-Strauss, Algirdas Julien Greimeas writes: 'Science originates from defining of its subject and from working out its methods not from the material that is randomly collected for it' (Greimas 1993:15). In other words, looking at contemporary Lithuanian ethnology what one will notice is that its material consists of 'mere' descriptions of rural traditions and the archive and of museum collections that have been accumulated as the 'Lithuanian ethnographic material'.

Greimas emphasises that such 'an ethnographic archive, organised by the independent efforts of countrymen is thereby heterogenic and can serve several ideological tendencies' (*ibid*). What ideological trends had challenged and tried to monopolise the field of ethnological research? Could Lithuanian ethnology, especially during the Soviet period, secure its profile as a scholarly discipline? What was the methodological impact in terms of meta-theories? How has the role of the discipline changed during the post-communist period? The aim of this article is to investigate the impact of the most influential methodologies and scientific paradigms as well as dominant ideologies on Lithuanian ethnology from its very beginning in 1930s to date.

Lithuanian ethnology is on the edge of scholarship and public interest in creating a repository for identity politics in a New Europe vis-à-vis the European Union. The most constant methodological challenge to Lithuanian ethnology during the different periods of its existence was a powerful field of history and the paradigm of the past. Both institutionally (the leading ethnological institution in the country – Department of Ethnology is still a part of the Lithuanian Institute of History; the only journal of the field *Lietuvos Etnologija*, is published there too) and as a research avenue, Lithuanian ethnology used to depend upon conceptualizations of the past provided by the discipline of history. Those 'who own the past' usually provide meta-theories about it, including, as it was during the Soviet period, historical materialism based on unilinear evolutionism.

The main ideological frame, which has characterised the discipline for decades, was nationalism. Actually, it was from the foundation of ethnology in this country, later followed by the resistant vs. conformist' strategies of the Soviet period. While Nordic and German ethnologies (renamed European Ethnology by Sigurd Erixon in the 1930s) was strongly influenced by American cultural anthropology in the 1960s, and was aware of innovations in social theory and opened the field to anthropologists, East European ethnologies were focused instead on the exploration of 'their own' cultures.

For the last seven decades ethnology in Lithuania has been understood as folk or traditional culture and everyday life studies or as ethnic culture studies (*etnines kulturos studijos*). The later is still the most popular term to describe the field of expertise of ethnology and folklore studies in Lithuania. In many cases it is a substitute or synonym for ethnology, as a scholarship that focuses on the study of a culture which is locked in ethnicity. In this case the culture is supposed to be Lithuanian and its bearers are supposed to be ethnic Lithuanians. Such an understanding is rooted in the identity politics of the interwar period.

Ethnology at the Universities of Vilnius and Kaunas

Cezaria Baudouin de Courtenay-Ehrenkreutzowa (1885–1967) was the founder of the first Institute and the Program of Ethnology established in Lithuania at Vilnius University, in 1924. The Institute has followed a descriptive/observational methodology, including the development of questionnaires, focused on field research and museology. The Ethnographic museum was established as part of the Institute as well. In the 1930s a well-known Polish ethnologist, Kazimierz Moszynski (1887–1959), brought from Krakow evolutionist methodology that was quite out-of-date at the time. However, according to his student, Prane Dundulienė, who later became a leading professor of Ethnology at Vilnius University in the Soviet period, 'after combining the methodological concepts of evolutionism and diffusionism, a trend was created that was called critical evolutionism' (Dundulienė 1978: 60–61). Nevertheless, Moszynski remained an advocate of an evolutionary approach to the study of world cultures. He supported this position despite the influence of diffusionism and certain cultural-historical or cross-cultural comparison trends that were already dominant in the Ethnology of Central Europe and the United States. His evolutionary stance was well represented in the volumes of 'The Folk Culture of the Slavs', written in Vilnius. The Polish period of Lithuanian Ethnology at Vilnius University, in general, left behind primarily ethnographic field descriptions and visual and artefact materials, particularly on the Vilnius region, that still remains on display in museum exhibits in the country.

Jonas Balys (b.1909) the most prominent Lithuanian ethnologist and folklorist of the interwar period, later of the Lithuanian diaspora in the U.S., made a major step in developing ethnology in Kaunas, the then capital city. He founded the Program and the Department of Ethnology (Etnikos katedra) in 1934 at Vytautas Magnus University and a year later the Lithuanian Folklore Archive, which became the leading centre for Lithuanian folklore studies with the international journal *Tautosakos Darbai* (Folklore Studies) edited by him. Balys saw Etnika first of all as folk/nation studies (*tautotyra*), it could be considered equivalent to German *Volkskunde* or Swedish *folklivforskning*. As he wrote,

> The very name [*etnika*] already shows that it is a science about the peoples (*tautamokslis*), the part of *etnika* that studies ourselves and our closest neighbours is *tautotyra* [folk/nation studies]. *Tautotyra* aspires to provide a real picture of the life of the European peoples, to understand the external and internal essence of every nation in its historical development. (Balys 1934: 16)

Balys was educated in and became a typical practitioner of Central–North European *Volkskunde*. In 1932 he defended his doctorate at the University of Vienna, at the school of Wilhelm Schmidt, the founder of *Kulturgeschichte* methodology in ethnology. Upon completion of his doctoral studies in Helsinki, where he received Habilitation for the book *Thunder God and Devil in the Balto–Scandinavian folklore*, and became a proponent of the Finnish historical– geographical method. When he returned to Lithuania, he gained a reputation as an active opponent of romanticism and evolutionism, two nationally prevailing paradigms in the field.

Balys tried to lay the basis for scientific accuracy in Lithuanian *etnika*, but he was criticised by his colleague, the evolutionist Juozas Baldzius Baldauskas, for insufficiently materialist or even idealist interpretations of data. Responding to Baldauskas, he put significant stress on his cultural-historical perspective with a strong positivistic stance: 'As there are no 'iron laws' in the spiritual sphere (neither they are absolute in the technical sphere), we therefore have to follow the historical method by first, collecting facts, then evaluating them critically and only then making conclusions (cf. Ciubrinskas 1993: 303).

The polemics between the two methodologies of evolutionism and cultural-historicism could be considered as symptoms of the scientific maturity of the discipline in pre-war Kaunas University. The ethnologists representing different methodologies, evolutionists Moszynski and Baldauskas, and the cultural-historians led by Balys, who introduced positivism and comparative historical method, left their followers in the post-war, occupied by the Soviets, Lithuania.

Local-regional descriptivism and museology dominated Lithuanian ethnology of the 1930s and the early 1940s. The main Lithuanian ethnology journal *Gimtasai Krastas* (Mother-Land) represented only museological ethnology. Nevertheless, the methodical variety and the appearance of scholarly polemics have proved that Lithuanian Ethnology can be a scholarly discipline.

Salvage Ethnography

Ethnologies in Central-East Europe being 'nationally operated anthropologies' (Hann 2003, Johler 2005) inevitably depended on their respective cultures and dominant ideologies there. Nationalism went along with Romanticism and the discipline of ethnology was born as a child of the Romantic nation-state building ideology of Central/East Europe. It acted throughout most of the twentieth century, and entered into the twenty-first as one of the central disciplines for national identity formation, nation-state building, nationalist ideologies, and even post-communist national revivals of the New Europe. Nation-building and 'nation-sacralization' issues were dominant in the politics of national identities through the region as well. So the 'national ethnographies' (Hann 2003: 16) limited themselves to their own nation's folk-studies, as it happened in the German-speaking countries with *Volkskunde*, and were quite often tightly linked with nation-building and national identity politics processes in their countries.

National ethnographies in the interwar period, *volkskundliche* understanding meant an interest in studding a 'pristine' local/regional/folk 'culture found and described in rural hinterlands of the nation-states'. Such an ethnography focused on a cultural-historical paradigm and in many ways, as Ernest Gellner puts it, it appeared as a 'salvage operation' of memory cultures, rescuing the memorial pasts in order to build a 'normative image of the traditional folk culture':

> The interest of folklorists and ethnographers lay in the description, collection, study, preservation, and often exaltation of their national (peasant) cultures. This holds true particularly of the countries of the 'third time zone' of Europe, 'which presented the greatest problems from the viewpoint of the implementation of the nationalist principle of one culture, one state ... Many of the peasant cultures were not clearly endowed with a normative High Culture at all ... [As a consequence] ... nationalism began with ethnography, half descriptive, half normative, a kind of salvage operation [*emphasis added*] and cultural engineering combined' (Gellner 1996: 115–16).

Such a nation-building and national culture 'salvage engineering' had its niche in the shaping of *Etnika* during the interwar period. Ethnological studies focused particularly on Lithuania and research 'expeditions' have been conducted exclusively in the country. There was no space for multiculturalism yet, and the comparativism of Balys, Baldauskas and others was very limited.

Soviet Ethnography instead of *Tautotyra*

The Soviet occupation of Lithuania brought significant changes to the social sciences and the humanities. Disciplines such as sociology and history, not to mention religious studies and political science were converted to the ideological framework of communist Leninist-Marxist ideology and their 'bourgeois science' background was questioned. The term 'Etnografija' (Ethnography) was introduced in all the institutions that pursued the studies of Ethnology. Since 1932, ethnology in Soviet Russia (*narodovedenije*), as well as socio-cultural anthropology, had been proclaimed 'bourgeois pseudo-sciences' and their place was taken by ethnography (Bromlej 1989, Bondarenko and Korotayev 2003). The term itself has produced a labelling effect, as post-communist Russian anthropologists Bondarenko and Korotayev have pointed out:

> The very fact that the discipline was invariably called ethnography produced a 'labelling effect'. Indeed, it was mostly ethnography in its pristine sense – i.e. 'description of the peoples'– rather than socio-cultural anthropology. Most ethnographers mainly studied topics relating to material culture such as (ethnic) housing, food, clothing etc., in order to establish patterns of historic cultural evolution (ethnogenesis, etc.) (Bondarenko and Korotayev 2003: 235).

In Soviet Lithuania, ethnology was defined as 'a branch of history which studies the peculiarities and development of the material, social and spiritual cultures of the peoples' (Vysniauskaitė 1964: 9). Institutionally, it became a subfield of history. In 1945, the subdivision of Archaeology/Ethnography was introduced at the Institute of History of the Soviet Lithuanian Academy of Sciences, and in 1961, it led to the creation of a separate department of ethnography.

Thus, the post-war *Etnografija*, like the interwar *Tautotyra*, lacked independence as a scientific discipline. Before the war, it was part of Museum Studies, and after the war it became a small branch of History. Nevertheless, the Section of Ethnography at the Institute of History became the main ethnological research unit in the country, completely separating itself from both Museum and Folklore Studies.

The subsidiary role of ethnology against the master discipline history has been proven methodologically. Some arrogant 'pure' historians, working with written sources, considered ethnological studies merely 'old wives' tales'. Thus, to prove his/her professionalism, an ethnographer had to be trained as a historian first, s/he could not trust the data collected in the field but rather had to verify it on the basis of the 'historical' (written) sources. Moreover, one had to apply strictly Leninist-Marxist historicism, which, especially during the first post-war decades, was straightforward unilinear evolutionism, requiring researchers incorporate their results into evolutionary, almost organic stages, called 'social orders', to be passed by all the peoples of the world.

The officially accepted methodological framework was historical materialism based first of all on the Morganian-Marxian schema of evolutionism. Ethnology, aspiring to its own scientific peculiarity, at best, could use the Soviet fraction of the diffusionist and cultural-historic methodology – the concept of typological method based on the concept of economic-cultural types and historical-ethnographic areas defined by the Russian ethnologist Nikolaj Ceboksarov in the mid 1950s (Ceboksarovas & Ceboksarova 1977). It was a combination of this unilinear evolutionism and of the cultural-historical approach. 'Types' and 'areas' were understood as developmental stages determined by the mode of production:

> Economic-cultural types are understood as certain complexes of economic and cultural characteristics that, historically, are formed by the different peoples. These economic-cultural types are usually in clusters in which adjacent nations are in close stages of social economical development and their populations live in similar natural-geographic conditions. Cultural-economic types are always connected to the mode of production of every particular society, because the mode of production ultimately determines the relationship between people and environment in every historical epoch (Ceboksarovas, Ceboksarova 1977: 137).

Since the late 1970s, unilinear evolutionism, under the impact of cultural materialism, turned into positivism, but it remained within the theoretical framework of historical materialism. Still under the 'iron laws of the development of progress' Soviet Lithuanian ethnography focused its research on the contemporary 'socialist everyday life styles of the rural and urban inhabitancy of the Soviet Lithuania'. For the purpose, quantitative analysis and statistic methods were deployed, and this profile was given the name of ethno-sociology. The methodology of positivism (which prioritises facts over theories and the necessity to collect and stock empirical (ethnographic) data prior to applying theories) that was dominant at the time was reminiscent of the historical particularist school of Franz Boas in the United States.

However, the subject and method of research of the national ethnologists was the Ethnography of Lithuanians. Fist of all it was understood as the traditional culture of the Lithuanian people. Both in pre-Soviet *Tautotyra* and in Soviet *Etnografija* the term the people's culture meant first of the traditional rural economy, countryside inhabitants and/or Soviet collective farmers all of that being labelled under the term '*liaudies kultura*' (the people's folk culture). Such a concept of culture itself echoed the ideology of Romanticism of the nineteenth century and was also used as focal for national identity politics. However, such a substitute for national culture had been questioned already by the 'architects' of the Lithuanian nation-state at the end of the nineteenth century.

Ethnography as a field research method was also understood in a particular way. It was the *Volkskundian* understanding, which meant collections and descriptions, and it had nothing to do with the Malinowskian revolution that led to 'presentism'. The latter was assumed as being wholly social science methodology rather strange for salvage ethnographies busy with memorable pasts. Ethnography, thus, was regarded as a method of field research into the folk antiquities of the countryside to be revealed through visiting selected informants in rural areas. It was the study of 'domestic exotics' to be found in the rural hinterland of nation-states. In Lithuania it was attributed the name of 'ethnic culture studies' which was, and still is, a substitute for ethnology, and the bearers of the ethnic culture in this case are the Lithuanian nationals.

Ethnographic data collecting have been organised via expeditions – that is field trips on the spot spending days and sometimes weeks in the area under research. Such 'ethnography' focused on the culture history and in many ways it appeared as 'salvage ethnography of memory cultures'. It was busy to document cultures or traits and patterns of 'traditions' which were 'disappearing' or were expected to disappear in the near future. The general method in the field was description, the 'gathering of data' consisted of verbal statements (tape recorded) and interviews with seated informants based on questionnaires.

The outcome of *Volkskundian* or Soviet ethnography was the descriptions and accounts, treated as valid information blocks (data) to make generalisation after. These blocks were used almost without any filtering and interpretation of field-notes against comparative theory, but quite often it was done against contextual documentary materials.

So, ethnologists focused their attention on the descriptions, iconography and collection of antiquities or memorable pasts found in the countryside. These antiquities were all categorised under the rubric of 'folk traditions,' and, being declared as 'fast disappearing', they were treated as if in need of urgent 'salvage' (i.e. collecting).

Such a research strategy actually continued the ethnographic-museological tradition of the pre-war period, and the need for collecting belonged to salvage ethnography of the nation-building period. Actually, the imperative of collecting fitted well with the dominant perspective of evolutionism, the central paradigm in the Soviet Lithuanian social sciences and humanities at the time. Evolutionism advocated the search for antiquities – relics or survivals in the present while holding, like in classic evolutionist E.B. Tylor's *Primitive Culture*, that these survivals were authentic fragments of the past of human development stages or Marxian 'social orders'. The Communist Party also obliged the ethnographers to undertake the recording/description of 'material culture and labour traditions that bring progress and produce boon' especially focusing on the studies of the 'socialist present' (Milius, 1992).

Finally, ethnology was primarily a conglomeration of ethnographic field studies which, as a result of its minute and thorough documentation on daily cultural materials, filled up archives, manuscript treasuries and museums with ethnographic collections. Ethnography provided an excellent impulse for publications of local-regional monographs on Lithuanian localities and built a strong empirical foundation for future scientific research.

Hence, the ethnography of the time was under the spell of positivist empiricism and was oriented toward collection. 'Until there is still material to collect' and 'until all the traditions have been collected [i.e., registered and archived]' their verification will be often postponed or replaced by an ideological assessment, both dominant and resistant.

Socialist *vs.* Nationalist Manipulation of 'Tradition'

Ethnology of the Soviet period has functioned more or less as an applied discipline serving two ideologies: counter-establishment nationalism and the dominant ideology that required an active participation of ethnologists (ethnographers) in the 'creation of new, socialistic traditions.' Since the general ideological line of the regime defined Soviet culture as 'socialistic in content and national in form,' the 'forms' could be national (sanctioned, of course, 'from the top') whereas the content had to be new, socialist and anti-religious. The creation of the new-socialist ceremonials of the family circle, for example, 'inventing' socialist and anti-religious baptising of children, marriage and funeral employed a certain dose of 'national forms', i.e. Lithuanian tradition. By all means, there was a strict selection of only particular forms, suitable for 'reinventing tradition', whose content had passed the muster with the Party guardians and would impeccably suit the Soviet nomenclature. Ethnology was handy for the dominant ideology as a provider of knowledge of the 'local forms' needed for the remodelling of the old, Catholic, and the invention of the new socialist traditions, i.e. celebration of the presentation of the Soviet passport, or Women's Day.

For ethnologists of the period to be involved in applied ethnology at the museum or folklore archive and to manipulate the logo of *tradicine liaudies kultura* (traditional culture of labouring masses) as a substitute for 'Lithuanian folk culture' was not complicated. Of course, more sophistication was needed to explain religious motives within that culture as either aesthetic or somehow 'politically correct'. It was much more complicated to get scholarly work published through the censorship and maintain your status as a scholar. It was necessary to follow the methodology of historical materialism eventually refurbished into positivism. Ethnologists therefore paid tribute to the creation of Soviet traditions by offering scientific advice for the standardization of the old cultural forms while also adapting to new contents.

Culture Collecting for Identity Politics

An open expression of the Lithuanian national identity politics during the communist regime was its provocative posture. It was backed up by romantic perspectives towards the nation's past and by patriotism. Some of the researchers were especially influenced by the nineteenth century romantic interpretation of the Lithuanian past. Paradoxically during the Soviet period, a high value was given to the spirit of the nation, and the ethnographic data revealing magic, mythology, ritual and symbols became assumed again as the manifestations of that. Despite scientific criticism, the recorded memoirs of a Soviet collective farm (*kolhoz*) 'peasant' might be seen as an echo of archaic traditions, and the patterns of homemade cloths found on a Soviet collective farm might be interpreted as authentic symbols of the ancient Balts.

'Culture collecting' for the Lithuanian ethnologists meant a truly patriotic mission of collecting authentic and typical survivals of the Lithuanian folk antiquities. This truly patriotic passion led to the filling up of museum collections and archives' manuscript treasuries. At the same time, however, it led to a certain casualty and even peculiar dissociation from ethnology, as researchers satisfied themselves exclusively with ethnography in the full sense of the word. Their main moral imperative of the ethnographies throughout the field motto was: 'Folk antiquities are disappearing, old folk tradition performers are dying out, therefore there is an urgent need for the salvaging of what is left *in situ*; the analytic research meanwhile can be postponed for the following generations' (Milius 1999).

As James Clifford maintains, culture collecting is a 'contested encoding of past' (Clifford 1988: 218), it serves for identity needs and from such an essentialist perspective cultures become ethnographic collections:

> Cultures 'are ethnographic collections ... Collecting ... implies a rescue of phenomena from inevitable historical decay or loss. The collection contains what 'deserves' to be kept, remembered, and treasured. Artefacts and customs are saved out of time. Anthropological culture collectors have typically gathered what seems 'traditional' – what by definition is opposed to modernity ...What is hybrid ... has been less commonly collected and presented as a system of authenticity (Clifford 1988: 215–31).

Following the imperative to collect the 'antiquities of their culture', Lithuanian ethnologists sought to find traces of 'truly traditional' culture and in searching for 'authentic traditions', they ended up with the pre-Christian Baltic past. It was this form of cultural fundamentalism backed also by mass interest in folklore that appeared in the 1970s and the 1980s throughout the East Europe (Skrodenis). Many folklore ensembles and clubs as well as local history studies clubs were founded as a result of the neo-Romantic nationalism wave enabled by Stalin death and the Khrushchev reforms in the 1960s. The engagement into folklore was in particular widespread among the younger generations. The new post-Stalin generation, mostly students, became fed up with the staged fake-lore of the Soviet establishment and having been inspired by neo-

Romantic zeal they developed an interest in performing 'authentic Lithuanian' folklore and even studding the culture of the ancient Baltic tribes (Ciubrinskas 2000).

Jonas Trinkunas, the head of the local history and folklore club in Vilnius, describes the importance of folklore during the period: 'Giving a new lease of life to the folklore, archaic traditions and festivals became a major task [for Lithuanians]. Youths would gather on a regular basis to learn and sing folk songs; traditions and mythology were being studied, archaic calendars and family festivals were being celebrated, mounds were visited and cleaned as well as other historical places … It was 1968–1969, the beginning of the mass movement of [folklore] ensembles. The get-togethers and concerts of folk music would become a real festival-halls would be packed. The joy wasn't exclusively aesthetic; a free spiritual territory that was not subordinate to anyone was discovered. In the get-togethers everyone sang and everyone danced – the boundary between the performers and their audience disappeared and it testified a successful return to the traditional folk culture. Requirement for professionalism was in second place, most important was authenticity, and the songs of the people of the own region as well as the mood of nationhood covering everybody' (Trinkunas 1996: 65–66).

From the perspective of both leaders and participants of folklore, ethnologists and folklorists were expected to perpetuate the Lithuanian traditions by acting as compilers of the collections of the 'culturally specific' even 'ethnically unique' and authentic pieces of the ancient Baltic past. They have been looked upon as potential creatures of the standard traditions too, which would suit for the practices of the, 'perpetuation' and 'revival' of the deprived Lithuanian culture. In this sense, the popularity of the books of Vilnius University professor Dunduliene and folklorist-mythologist Norbertas Velius gained previously unseen popularity.

As experts of traditions, ethnologists and folklorists were welcome in various folklore, history and nature fun club meetings that were often on the verge of legality and illegality, and that were mostly neo-Romantically oriented. They gave lectures and consultations for the local history studies and tourist clubs, and in particular to the folklore ensembles, where the need for 'authentic folklore' was high and the peasants' traditions were especially valued.

Thus, ethnologists played a significant role in anti-Soviet nationalist identity politics by striving to document the layers of typical and specific authenticity of the Lithuanian traditions in Soviet Lithuanian present, sometimes even to reconstruct the pre- Christian-, Lithuanian-, Baltic-, and Indo-European heritage. Though archaisms may not be found, there remained a need to describe them. An example of this is the numerous descriptions of probably random markings on stones as Baltic religious symbols (*Dunduliene*).

Such an essentialist 'culture collecting' documentarism served as counter-establishment research strategy of the period. This research focus was counter-establishment, ideologically anti-Soviet and against Party imposed ideological guidelines to study only the overt materialist-cultural traditions. This formulation of a distant romantic past predating the Soviet present excluded Soviet dogma in theory and method and increased the importance of the role of applied ethnology (or ethnography) as well as Folklore Studies. Ethnographers were 'in competition' to find 'the most archaic past.' For this reason, many ethnologists would sometimes trespass the boundaries of the positivist analysis of traditional folk culture and even the boundaries of any academic discourse to reconstruct 'deep cultural layers'. They would claim to have found fresh traces of the primeval ethnic Lithuanian culture. They would also attempt, like contemporary history studies, to record or 'fixate' the 'a priori view of the past ... and be the guardians of a collective memory ... involved in actions aimed at protecting the nation' (Suziedelis 1996: 18).

There is no doubt that during the Soviet period studies of history were endangered through their politicization and were thus turned into a sacral-mystical version of memory that is resistant to critics (Suziedelis 1996: 14). Ethnology was also threatened by being turned into a tool for scientific argumentation for the assimilation and/or denationalization of the Lithuanians into the melting pot of the Soviet Russian Empire. Especially during the mid-1960s, when there was an urge to implement the Brezhnev ideology of the 'fusion of nations'. Not only did they follow the methodology of historical materialism, but also showed that the lifestyle and consumption of Lithuanians who worked in towns and in collective farms (*kolkhoz*) were essentially the same when compared to workers from other 'brotherly republic', as, for instance, Tadzhikistan in Central Asia. Ethnologists reacted to this by being reluctant to create the new Soviet socialist traditions as well as by striving to save 'archaic traditions' for the possibility of their detailed scientific recording.

There was an anti-establishment and national identity politics based, if not a nationalist drive to make ethnology (as well as folkloristic) a nationally relevant discipline for nurture of the relicts of Lithuanian traditional culture. It was assumed that the subject of research itself as well as the analytical experience of the folk/ national traditions implied their relevance. A professional promise was to perpetuate Lithuanian ethnography collections as authentic texts or as compendia of standard national traditions. While being 'scientifically valid,' such compendia were suitable for local practitioners', like local history, environment clubs and folklore ensembles in their activities of the traditional culture retaining and revival.

In any case, there is no doubt that the ethnologically-documented archaic and therefore 'authentic Lithuanian' traditions, in ethnologic terms, 'typical and specific local or regional cultural traits' were an effective symbolic power against the modern, and therefore 'inauthentic, by the establishment created, Soviet culture brought into Lithuania from Russia. Anti-establishmentalism reveals itself in putting stress on the pre-soviet past and ignoring the Soviet present. Nevertheless the processes of invention of tradition brought together the establishmentalists and their opponents.

In its concern for folk culture and traditions Lithuanian nationalist ideology of the 1960s–1980s was akin to the Swedish Romantic intelligentsia – 'culture builders' laying cultural background for nation-state at the end of the nineteenth and the beginning of the twentieth century (Löfgren and Frykman 1987). The Herderian concept of *Volksgeist*, the 'genuine folk spirit' enshrined in traditional culture and language, was invoked during both periods. The Romantic intellectuals of the nineteenth-century Europe and the Lithuanian neo-Romantic revivalists of the 1960s–80s shared the same general attitude to the folk culture, identifying it as the 'peasants' culture' as 'genuine', 'pristine', 'ancient' and 'traditional' and, eventually, 'national' at the core (Ciubrinskas 2000: 42).

The movement of the animators of the national traditions, which was resistant to the dominant Soviet ideology in the 1960s through the 1980s, was essentially neo-Romantic. From the revivalist point of view, the ethnographic and folklore collections were viewed as if they were a repository or treasury of the nation, and the ethnographers aspired to fill this treasury with their own items collected under the same romantic zeal. Those who were engaged in ethnographic studies were considered to be at least potentially critical and dissident in respect to the dominant ideological-political conjuncture. Those engaged in the local history studies as well as the folklore ensemble performers and organisers were not as much expected to have scientific expertise as they were expected to reflect the national spirit that would inspire retaining of traditions and offering models for their revival. Therefore, the lectures and books of the leading folk culture experts of the period Velius and Dunduliene who interpreted Lithuanian rituals, myths, magic and symbols became very popular.

The impact of the ethnologists as scholars depended upon their professional skills (the skills of 'collecting, but also their use of primary sources for the 'reconstructions' of the 'authentic traditions'). Previously unreported traditions were 'brought back to life' and there was an effort if not to revive then at least to celebrate public festivals according to the data obtained. In Kernave, the pre–Christian summer solstice ritual festivals were revived since 1967, and performed every midsummer until today. The festival serves as an eloquent example of the reconstruction of previously unreported traditions. The Soviet regime attempted to persecute members of this ethnological revival: 'despite the persecution and disposition by KGB, this [folklore] movement and especially St. John's day celebrated in Kernave was viewed with suspicion ... There was a deep interest not only in the traditions and archaic lifestyle but also in pagan beliefs, putting their reconstructions in opposition to militant atheism (Apanavicius 1996: 59). The two ideologies, competed with one another, the dominant one, which worked to create a socialist tradition in opposition to a nationalist thrust, full of neo-Romantic zeal, striving to recreate the 'the golden past' and to secure a status quo for survivals of the traditional culture. This was particularly evident in the applied field of the discipline, where ethnologists worked as experts of traditions.

The Ethno-culturalism of Ethnology during the Perestroika and post-Communist Period

The neo-Romantic zeal of the 1960s and 1970s was undertaken by the creative intelligentsia of 1980s and fitted well into the process of democratisation started during the Perestroika. The National identity politics embraced the idea of the Lithuanian national culture being rooted in folk-culture, even in the ethnic culture of the Lithuanians and the ancient Balts, as it was argued by the most influential ethnologist of the period Velius (Velius – 1991 – *Liaudies kultura*). The search for the genuine folk culture, renamed by Velius into ethnic culture received mass approval at the end of the 1980s perestroika, followed by the Singing Revolution.[1]

During the period of perestroika of the late 1980s there was a quest to restrict ethnology to traditional cultural study, or ethnic culture (*etninė kultura*), that is the discipline that dealt with the 'most authentic' oldest and deepest roots of the Lithuanian culture. 'Ethnic culture studies' (*etnines kulturos studijos*) coined by Velius, is still the most popular terminology with which the field of expertise of ethnology and folklore is described in Lithuania nowadays. In many cases it is a substitute for ethnology, as a scholarship which focuses on studying the culture that is locked in ethnicity.

At that time, the term *etninė kultura* held a threefold meaning: firstly, it indicated Lithuanian traditional folk culture; secondly, it was the subject of the discipline of Lithuanian ethnology; and thirdly, it referred to 'the traditions' to be retained and reconstructed by professional ethnologists. So it was high time for ethnologists, and there were numerous offers from the increasingly media to write articles or speak about 'ethnic culture' (i.e. the ancient Lithuanian mythology, rituals, ceremonies, traditions, festivals etc.).

After independence in 1991, and during the re-establishment of public life in Lithuania, models of Lithuanian ethnic culture dominated public representations of culture. This was particularly explicit in public and private discourses about 'true' Lithuanian identity when the network of 'ethnic culture' institutions mushroomed in the country.

The discipline of ethnology was renamed into 'ethnic culture studies', and as a synonym for traditional Lithuanian folk culture it was publicly approved in the new post-Soviet school manuals: 'the new Lithuanian school faces the important task of nurturing ethnic culture, of encouraging the recognition of its expressions by schoolchildren, to teach them values and perpetuate the traditions of ethnic culture.' (Cepienė 1992: 3)

The Singing Revolution has showed the effectiveness of Lithuanian ethnology as an applied discipline as well as a scholarship that had resisted the Soviet regime. The Soviet and the early post-Soviet politics of the discipline have influenced the contemporary emergence of Social Anthropology, a different discipline with a comparative relativistic reach, a holistic concept of culture, not binding the observer to his or her roots or traditions.

1 On 23 August 1989, fifty years after the Molotov–Ribbentrop pact about the Nazi-Soviet occupation of the Baltic countries was signed, a human chain of joined hands from Vilnius to Tallinn via Riga was organised by the National Fronts of all free countries. The so-called 'Baltic Way' was the beginning of the Singing Revolution in the Baltic Countries, which drew upon a hundred-year tradition of mass singing, and took place during annual National Folk Song Festivals that were organised regularly since the end of nineteenth century. In 1990, Sajudis backed candidates who won the elections to the Lithuanian Supreme Soviet and obtained Lithuanian independence, with Lithuania becoming the first of the Soviet republics to declare separation from the USSR.

Anthropologisation of Ethnology and the University Politics

In the era of post-communism a few East-Central European ethnological (former ethnographic) institutions changed their names to ethnology and cultural anthropology. The new labels recognise the fact that anthropology has become fashionable along with other trends in Western scholarship. Folklorists and ethnographers gave up their identities overnight, and, called themselves 'anthropologists'. Former departments of ethnology (or ethnography, Volkskunde etc.) were now named departments of ethnology and cultural anthropology (Godina 2002:13).

The tendency to equate ethnology with anthropology according to the Slovenian anthropologist Vesna Godina is directly linked to the 'money and power dilemma' (Godina 2002:9). It became both fashionable and fruitful to use the label of anthropology to attract prestige and raise funds. New establishments, departments, programs etc. provide a rare opportunity to attract new funding for research and teaching, as well as providing new power bases in the positions of deans, heads and chairs.

This is particularly true of Vilnius University where the Program of Cultural History and Anthropology was launched in 2000. It was supposed that anthropology would take a small niche previously allocated for ethnology, as a branch of history. However, interest in the anthropology courses grew if compared to history. After the Program became popular among students, due to the University politics, it was reshaped so as to exclude almost all courses in anthropology and leave only the name of anthropology its title. Since then, anthropology courses have no longer been offered, and the School of History has continued to attract applications from students because of the name of the Program. Thus, the field of history at Vilnius University has gained new resources though it has cared little for either anthropology or ethnology.

I agree with Peter Skalnik (Skalnik 2002) and Vintila Michailesku (Michailesku 2005) that the discipline of anthropology does not appear as distinct from ethnology (see Skalnik 2002). Eastern European countries retain a clear difference between ethnology, with its long and recognised tradition, and socio-cultural anthropology which lacks, in most cases, any tradition at all (Godina 2002: 7). Structure and personnel, not to mention methods (ethnographies), have to be different. Gathering, collecting, describing and registering expeditions of ethnologists and long term fieldwork which includes participant observation, done by anthropologists is different not only in the amount of time spent in the field but also in content, orientation and research paradigms such as synchronic investigation and comparative methods.

Skalnik points out that there are significant differences between ethnology and anthropology: 'Those who maintain that ethnology (ethnography) is a synonym for anthropology and therefore anthropology is not actually needed, underestimate the strength of the historical sciences tradition, for they must know well that by making no distinction they automatically – in the specific conditions of Central-East Europe – help to preserve the ancient regime (i.e. the Soviet system) (Skalnik 2002: VII–VIII).

Chris Hann has described the process of contemporary European Ethnology becoming national anthropology (Hann 2003), which is synonymous to 'anthropology at home'. Its practitioners in Lithuania subscribe to the thesis that ethnology can be equated with social and cultural anthropology. However, Godina points out that delimiting ethnology from anthropology rests on mostly unclear criteria. The modern paradigms of 'anthropology at home', 'anthropology back home', a 'native anthropology' or 'indigenous ethnology' were not recognised by the Central/East European ethnologists (Godina 2002). The East European ethnologists' point of view is that, in so far as West European anthropology no longer insists on otherness as its subject of study, their approach is socio-cultural and always has been. The praxis in post-socialist countries, therefore, is marked by a tendency to equate 'reclassified' ethnology with anthropology (Godina 2002: 9). Such a unifying or interdisciplinary hybrid direction took shape in Lithuania as well and the journal *Lithuanian Ethnology: Studies in Social Anthropology and Ethnology* was founded in 2001. Editors of the journal aim at exploring new ways of integration of integrating regional ethnology and social anthropology.

Cosmopolitan Anthropology's Challenge to National Ethnology

Anthropology is about the other; it deals with otherness. This may sound as 'a contradiction in terms' (Hastrup 1993: 155, 161, from Geana in Skalnik 2002) since East Europeans are still concentrated more on romantic and nativist intellectual tradition exploring their archaic culture, describing its reminiscences and artefacts. It is only recently that East European anthropology has started to discuss the distinction in between 'we' and 'the other'.

Conformity, xenophobic and arrogant attitudes towards anthropology are predominant in Lithuania. These are rooted in the general perception that 'culture' is an intellectual achievement and/or confusion with national 'ethnic culture' (Vastokas 2005). In 1990, the Department of Anthropology at the Vytautas Magnus University in Kaunas was founded with the support from the Lithuanian diaspora in the US and Canada. The head of the department and three other anthropologists of Lithuanian background from the US and Canada started to give instruction in cultural anthropology. Before the program in anthropology was fully established, however, after two years of operation, the Department was re-structured (Anglickiene 2001) and 'integrated' into the newly-formed, but actually old fashioned, *Volkskunde* focused department of Ethnology and Folklore Studies. It was a step forward towards conformity with the predominantly 'nationalist' educational politics in the country. A highly-placed academic commented on the decision suggesting that 'we don't need to be taught about Africa: there is an urgent need to learn about our traditions instead. Even more so, we should learn more about our traditions because they are dying and the former, Soviet regime was not in favour of studying it' (Sauka 1999).

An example of xenophobic attitude comes from a round-table discussion held by the George Soros Foundation in Vilnius in 1999 entitled 'Does Lithuania need Socio-cultural Anthropology?' The question raised by the moderator, 'Couldn't Lithuanian ethnologists do what anthropologists do?' implied the equivalence between a home-bred ethnology and a 'cosmopolitan' anthropology. Socio-cultural anthropology was intended as a scholarship with no 'tradition in the spectrum of national science development'. One of the participants, the Director of the leading folklore research institution in the country even asked: 'Do we really need this novelty? Are we not capable of achieving these proposed aims within existing research fields and institutions and within existing resources and researches?' (Sauka 1999). These attitudes are by no means universal, but they provide cash-starved university administrators with ready-made arguments on why another department, centre, or program should not be set up, which would duplicate potential analogues already in existence.

The nationalist ideology and politics of rebuilding the nation-state, in the early 1990s, promoted 'national' disciplines – Lithuanian history, language, folklore by reinforcing their role as 'identity cornerstones' while dismissing 'globalizing' ones, like anthropology, for 'not being in focus'. 'Home-bred ethnology' against 'cosmopolitan' anthropology enables the 'national', and well established disciplines to monopolise some basic social science categories such as 'ethnicity' or 'culture' by including them in the curriculum of the newly emerged discipline of 'Lithuanian ethnic culture studies' at Vilnius Pedagogical Institute and teaching courses such as 'Cultural anthropology of Lithuanians' at Vytautas Magnus University in Kaunas.

Conclusion

The path of Lithuanian Ethnology from its beginning in the 1930s to date shows that ethnology was and still is a field of great significance influenced by the politics of identity. Already in the 1930s, and especially during the Soviet period, it was marked as 'salvage ethnography'. Methodologies have changed, from the evolutionism to *Kulturgeschichte*, to historical materialism and eventually positivism. Similarly, ideologies have changed from nation-state building nationalism of the interwar to confrontation of communist *vs.* nationalist and eventually ethno-nationalist, after the fall of the Soviet empire. What remained unaltered throughout those periods were the public and the professional ethnologist interest in management and manipulation of folk culture traditions for the re-conceptualization of national identity. These remain significant features of the post-communist Eastern Europe: as Hann notes, in fact, not even the experience of socialism, with its ostensibly universalistic and internationalist ideology, could lead to a radical break with the habits and research practices of national traditions (Hann 2003).

During the Soviet period, the discipline of ethnology in Lithuania risked to become a tool for the dominant ideology as a resource of scientific argumentation for the assimilation of the Lithuanians into the melting pot of the Soviet Russian Empire. At least it had to follow the methodology of historical materialism. The field was also marked by some non-conformist attempts. The neo-Romantic approach to folk culture was revived and ethnology became again a resource for salvage operations against the Soviet establishment. Its methodological paradigm was the search for authenticity. Thoroughly well explored by positivism, it enabled Lithuanian scholars to confront the Soviet establishment. The 'authentic Lithuanian' traditions that were called by ethnologists 'typical and specific local or regional cultural traits' were an effective counterbalance to the new, and therefore 'inauthentic', Soviet culture that was brought into Lithuania. It also allowed for the archaicising project of modernity by securing status quo for the survivals of the traditional culture. This was particularly seen in the applied field of the discipline practiced by ethnologists who worked as experts in the manipulation of the folk traditions.

Bibliography

Anglickiene, L. and Senvaityte, D. (2001) 'Etnines kulturos vagos gileja', *Darbai ir Dienos*, 25, 292–94.

Bondarenko, D. and Korotayev, A. (2003) 'In Search of a New Academic Profile', in D. Drackle, I. Edgar, and T. Schippers (eds), *Educational Histories of European Social Anthropology*. New York, Oxford: Berghahn Books, pp. 230–46.

Bromlej, J.V. (1989) 'Oktiabr i etnologiceskoje izucenije sovremenosti', in *Etnokulturnyje tradiciji i sovremenost*. Vilnius: Lietuvos TSR MA Istorijos institutas, pp. 5–30.

Cepiene, I. (1992) *Lietuviu etnines kulturos istorija*. Kaunas: Sviesa.

Ciubrinskas, V. (1993) 'Afterword', in Balys, J. (ed.), *Lietuviu kalendorines sventes*. Vilnius: Mintis, pp. 302–07.

—— (2001) 'Challenges to Lithuanian Ethnology during the Soviet Period: the Discipline, Ideology, and Patriotism', *Lietuvos etnologija: socialines antropologijos ir etnologijos studijos*, 1(10), 99–117.

—— (2005) 'The First Program in Anthropology in the Baltic States at Vytautas Magnus University in Kaunas, Lithuania', *EASA Newsletter*, 39, 6–10.

Dunduliene, P. (1978) *Etnografijos mokslas Vilniaus universitete*. Vilnius: Lietuvos TSR aukstojo ir spec.vidurinio mokslo ministerija.

Frykman, J. and Löfgren, O. (1987) *Culture Builders*. New Brunswick, London: Rutgers University Press.

Gellner, E. (ed.) (1980) *Soviet and Western Anthropology*. London: Duckworth.

—— (1988) *State and Society in Soviet Thought*. London: Blackwell.

Godina, V. (2002) 'From Ethnology to Anthropology and Back again: Negotiating the Boundaries of Ethnology and Anthropology in Post-Socialist European Countries' in P. Skalnik (ed.), *A Post-Communist Millennium: the Struggles for Sociocultural Anthropology in Central and Eastern Europe*. Prague: Set Out, pp. 1–22.

Greimas, A.J. (1990) *Tautos atminties beieskant: Apie dievus ir zmones*. Vilnius and Chicago: Mokslas.

Hann, C. (2003a) 'The Anthropology of Eurasia in Eurasia', *Max Planck Institute for Social Anthropology Working Papers*, 57. Halle/Saale.

—— (2003b), 'Three Levels of Cosmopolitanism in Anthropology', *The First Baltic States Anthropology Conference Books of Abstracts*. Vilnius: Vilnius University Press, pp. 11–12.

Harris, M. (2003) *Kulturine antropologija*. Vilnius: Eugrimas.

Hohnen, P. (2003) *A Market out of Place? Remaking Economic, Social and Symbolic Boundaries in Post-Communist Lithuania*. Oxford: Oxford University Press.

Milius, V. (1992) 'Etnografijos pasiekimai ir rupesciai', *Liaudies kultura*, 2(23), 10–23.

—— (1999) *Presentation at the Conference of the Lithuanian ethnologists held in the National Museum of Lithuania.*
Sauka, L. (1999) *Presentation in the discussion organised by Open Society Lithuania on the Place Anthropology of in Lithuania.*
Skalnik, P. (ed.) (2002) *A Post-Communist Millenium: The Struggles for Sociocultural Anthropology in Central and Eastern Europe.* Prague: Set Out.
Vastokas, R. (2005) 'From Glasnost to NATO: Retired and Restless in a Post-Soviet State', *Trent University Newsletters*, 2, 14–16.

Chapter 6

When is Small Beautiful? The Transformations of Swedish Ethnology

Orvar Löfgren

In a discussion on the life of disciplines Marjorie Garber talks about 'disciplinary libido', the ways in which a field differentiates itself from other disciplines, but also may have open or secret desires to become like its nearest neighbour (Garber 2001: ix). What would the libidinous history of European Ethnology in Sweden look like? What drives and energies have developed and how have they been channelled?

Within a Scandinavian context, disciplines like ethnology, folklore, history and archaeology were part of a cultural nation building, albeit in different ways, carried out by small nations on the periphery of Europe. Once upon a time, the kingdoms of Denmark, Norway, Sweden, and Finland were organized as aggressive maritime mini empires with a great appetite for military conquest. While Denmark became more oriented towards the North Atlantic, Sweden looked towards the Baltic. By the mid-nineteenth century, these small states had lost most of their colonies or earlier mixed ethnic territories, and turned their gaze inwards, and sometimes backwards to a more glorious past and a true folk heritage. They now redefined themselves as small, peaceful nations with relatively homogeneous populations contained within 'natural borders'.

While many European nations with strong colonial traditions or imperial ambitions at that time started to create a global kind of anthropology, small nations like the Scandinavian ones turned to discover 'their primitives within', either in the form of folklore studies or as a more general ethnology of the nation.

In Scandinavia there was a parallel development of ethnology (often called folk-life studies) and folklore during the latter part of the nineteenth century and the early twentieth century, but with some interesting variations. In Denmark archaeology occupied the role of the most prestigious discipline of the national heritage; here ethnology as an academic field developed much later. Norway, up to 1905, joined in a forced union with Sweden, both folklore and folk-life studies became part of the claim to the uniqueness of the Norwegian national heritage. Finland, lost by Sweden to Russia in 1809, enjoyed considerable cultural and political freedom in the Russian empire and here cultural nation building was intensified up until independence in 1917 producing a very strong folklorist tradition, while 'Fenno Ugric' ethnology benefited from its position within the Russian empire, and developed a much more transnational and broader approach than other Nordic ethnologies (Bo Lönnqvist

2005). (Finns were often recruited as administrators, officers and explorers for this vast empire).

The focus of folk-life studies and folklore also revealed a basic structure of national culture building, the constant aim to contrast one's own heritage to that of one's neighbours or competitors. Thus, it is no surprise that the true Norwegian folk culture was found among the peasants in the remote mountain valleys of Telemark, or that the atypical and conservative province of Dalecarlia came to symbolize Sweden and that the true Finnish folk culture survived in the forests of Karelia.

Unlike European ethnology, social anthropology emerged rather late in Scandinavia. It lacked the support of a network of both central and regional museums as well as the moral support of cultural nationalism. Although general anthropology and European ethnology developed within the same traditions of cultural theory, and the early pioneers read much of the same classics, their position in Academia came to be very different. European ethnology and folklore was defined as belonging to the humanities with links to history, literature, art history and languages, whereas anthropology was initially seen as a natural science, with strong ties to geography and other natural sciences. This division of labour can be seen in the establishment of the national museums during the nineteenth century. In Sweden anthropology belonged to the Natural History Museum, European ethnology to the Nordic Museum, and there was a long fight about who had the right to the Sami minority (or as they then were called the Lapps). Were the Sami part of the Swedish national heritage, and thus part of the Nordic Museum, or should they be seen as an exotic tribe, which belonged with the other primitives of the zoology collections?

How significant is it that we, in the Nordic countries, ended up with a clear division of labour between a general anthropological perspective and a regional specialization (with a historical dimension) unlike many other European settings? From the end of the nineteenth century onwards, a new discipline staked out its territory in these countries, thereby shaping not only its own identity but also the orientation and aims of neighbouring subjects. In countries with no tradition of European ethnology, the field of cultural studies was divided in a very different way.

Small is Beautiful

In 1918, the first Swedish professor of folk-life studies, Nils Lithberg, held his inaugural lecture at the Nordic Museum in Stockholm. His chair was named 'Nordic and comparative folk-life research', and his presentation of the new academic discipline was comparative. He discussed how Swedish ethnologists should relate to international research in ethnology and cultural history, and in his discussion he moved quickly between different continents and eras. Why is the mentality of a European different from that of a Hindu? How is the use of burial trees in Dalarna related to similar traditions among Austrian peasants? He ended on a grand note, stating that European ethnology is the study of Man and that our task is to find the answers to mysteries of the human mind. In Lithberg's generation, and especially among his folklorist colleagues, we find this grand, comparative approach and a close link to the contemporary and general anthropological theories of evolution and diffusion, which made research both comparative and international, but also rather speculative.

There was, also, another relevant aspect. In 1918, Lithberg also published a study of pepper mills among Swedish peasants. He started with a general statement of the role of the ethnologist:

> It is not enough to take an interest solely in the things that strike the eye because they look curious. A collecting project, if it is to satisfy scholarly demands, should not neglect a single tiny thing, no matter how commonplace it may seem, among all the many things that belong to the everyday surroundings of the people. (1918: 19–20)

The interest in 'tiny things' was due mainly to their role as trace elements in studies of distribution, in the evolutionist and diffusionist spirit of the times, but his text reveals an engaging concern for the small, seemingly trivial and valueless things. Lithberg wrote about many such small objects: essays about cow bells, horn spoons and basketwork. Folklife scholars had been entrusted with the task of collecting and preserving everyday, overlooked objects: baking shovels, hoes, milking stools. For established colleagues in the old humanistic disciplines working with Art, History and Literature, this ambition could easily seem comical or trivial.

As a newcomer in the academic field, Swedish ethnology was seen as small in many ways, small in numbers, small in its scope, and small in its outlook. This reputation lasted until the early 1960s when, as a young student, I told my history professor that I wanted to switch to ethnology in order to be able to do cultural history. He answered sarcastically: 'Folk-life research, that's not a science, it is nothing but the study of peasant brawls.'

When I entered the Institute of Folk Life Research at Stockholm University I was just in time to witness the death of a long and ambitious project. From the 1920s to the 1960s it was the mapping of folk cultures that gave the discipline of folk-life studies its focus, structure and stability, channelling the academic libido into a common goal that would unite old professors and young students. The author of a small student paper would know that she or he at least contributed to the great project of the Swedish folk atlas, if only by providing a footnote.

It was Lithberg's successor, Sigurd Erixon, who launched the project of a Swedish Folk Atlas, which was part of an ambition to develop ethnology into a less speculative and more exact science. He was an extremely assiduous fieldworker who used to say that there was not a parish in Sweden that he had not visited, and he wrote just as assiduously.

The geographer, Anne Buttimer (1992), has analysed the life cycle changes of disciplines and pointed out that the pioneer era is often characterized by a phoenix period, when a new discipline emerges and tries its wings. In this early era there is a lot of experimentation and an open attitude. But the phoenix is usually replaced by Dr Faust, who enters as a disciplining force, whose task it is to transform the flights of the phoenix into a well-organized academic subject. Faust is busy structuring the discipline, inventing concepts and definitions, drawing up boundaries, organizing archives, conferences and writing handbooks. His work may seem more mundane than the happy flight of the phoenix, but it is a necessary stage. In Swedish ethnology Sigurd Erixon, among others, represented this phase.

Like other subject builders, he had a kind of Linnaean ambition to name and classify. He and his contemporaries indefatigably created scientific terminologies for everything from agricultural tools and farm structures to details of home furnishing. He believed in the urgency of rescuing the disappearing peasant culture, measuring dilapidated barns before they collapsed, studying villages before the last open fields were enclosed, collecting forgotten artefacts hiding in attics and outhouses. Every field had to be covered especially when it came to the material culture of peasant society.

As knowledge increased, ethnologists had to learn a growing body of facts, about corner jointing techniques in log buildings, about the shapes of ploughshares, about textile production and food preparation. The subject became one where empirical knowledge was primary, which meant that there was less time or place for discussing theory. It was important to carry out empirical studies, since there were still so many blank patches on the map of Swedish peasant culture. There was a constant feeling of urgency, traditional folk culture was seen as dying and ethnology became a discipline manned by salvage teams.

Erixon, just like his folklore colleague, Carl Wilhelm von Sydow in Lund, was a small-scale empire-builder, a great organizer and European entrepreneur. Both von Sydow and Erixon saw themselves as pioneers, always ready to export ideas, approaches and organizational forms to other new European departments of folklore and folk-life. (To some extent they were competitors, but in the end folklore studies disappeared as a separate discipline in Sweden, and were merged with ethnology, which had a much more stable infrastructure and job market in the forms of museums and other institutions). These 'mapping' projects allowed for cooperation at the Scandinavian, European, and global levels.

In the end, these grand projects developed into routine. The atlas project turned into an old ocean liner that kept moving forward even when the engines had burned out. When I started reading ethnology in the 1960s, the ocean liner was still there, but it was stranded. As young students we moved around in a landscape of ruins from the Sigurd Erixon research industry at the department in Stockholm. On the abandoned desks we found boxes of excerpts, half finished maps and long protocols of evidence collecting dust. We never had a chance to experience the enthusiasm and the exhilarating feeling that went with the idea of a common project uniting the discipline. For us much of the earlier knowledge was dead. We needed to develop a new utopian project. The same disillusion was found elsewhere on the European scene, but took rather different forms. In Germany the '*Abschied vom Volksleben*' of the 1960s was a much more dramatic revolt against the old generation. In Sweden the revolt lacked the political edge of the German historical situation with the need to scrutinize the Nazi past of the discipline. (Strikingly enough there has never been a thorough analysis of the politics of Swedish ethnology, after or before the Second World War.) Secondly, Swedish ethnology turned towards a totally different direction, when it came to finding new tools for reinventing the discipline.

In Search of the Microcosm

There was a radical shift in the mental world-map of Swedish scholars. German speaking *Volkskunde* disappeared, and was replaced by British, American and Norwegian social anthropology and to some extent qualitative sociology of the American and British brands.[1]

Thus, in Scandinavia and in Germany, there were parallel attempts to reinvent European ethnology in the 1960s, though with different results. While both managed to import a new social theory and a marked interest in contemporary culture, the ethnological research practice and theoretical profiles in Germany and Sweden during the 1970s was different from what it had been in the 1950s. Only in the 1990s did the similarities become striking again.

Thus, there are many reasons for such developments: in Germany, the '*Abschied vom Volksleben*' coincided with a strong development of critical theory in the spirit of the Frankfurt school. Inspiration came mainly from within Germany, from social theory and social philosophy. (This Frankfurt influence not only directed the choice of topics and questions, but also the style of research and presentation.)

In Sweden the situation was totally different. The Swedish '*Abschied vom Volksleben*' was not, as I have discussed elsewhere (see Ehn and Löfgren 1996) a child of 1968, but an earlier disillusionment with ethnological research. There was not much inspiration to be obtained locally from either history or sociology; instead an anthropologization of the discipline took place. With the dying energy of the Folk Atlas project, new utopian visions were needed to bolster the academic libido. The new utopian project was 'Discover Sweden', and the rallying cry was 'back to fieldwork'. By and by, the distance to what was seen as the era of 'classic peasant culture' had grown, ethnology had slowly turned into a discipline of reconstruction and archival work, searching for relics of a past folk culture.

The new conception of fieldwork mainly came to mean community studies. This new interest really dates back to the 1950s, when the American anthropologist Robert Redfield had visited Sweden, charismatically pleading for the study of 'the little community'. Inspired by him, several ethnologists went out in quest of this microcosm. A new conception of 'small as beautiful' developed.

In the 1960s, this interest in local communities grew in strength to become a dominant mode of thought. Those who received their education then learned to see Sweden in terms of local communities. If we look at the choice of student essays and dissertation topics in this period, we see the emergence of views of which communities were more community like than others. This created a new selection principle, which was influenced in large measure by contemporary anthropological theory, both the functionalist and the interactionist variety. Such an interest focused on the periphery of society rather than the mainstream. It is in this light that we should see the great interest, for example, for fishing hamlets, which represented the perfect cultural form of the little community: isolated, homogeneous, well integrated,

---

1 The swiftness of this shift is illustrated in doctoral dissertations from the late 1960s: over a couple of years nearly all German references disappeared and Anglo Saxon titles took over.

self sufficient, and so on.[2] The disproportionate number of studies of such marginal settings was a quest for communities that were as 'exotic' or 'anthropological' as possible. With this search profile, the study of working class settings was chiefly concentrated in small factory towns, and metropolitan studies focused on 'urban villages', such as traditional, close-knit neighbourhoods.

There was a paradox in this development: in many ways it felt like a liberating period of internationalization. We were all busy reading international anthropological theory, but on the other hand research became intensively Swedish. Compared to the perspective of diffusionist and culture area studies of earlier generations, our geographical space was narrowed. In retrospect, Erixon and von Sydow worked within much broader areas. The prefix 'European' became more of a rhetorical statement as very few Swedish ethnologists did their research outside Sweden in the 1960s and 1970s.

The interest in local communities also aspired to let the little community reflect society at large. The English anthropologist Ronald Frankenberg's classic study *Communities in Britain* (1966) was based on this idea. Here, an array of community types, from the little agricultural village to the city suburb, was threaded together to illustrate English society. The macrocosm became the sum of a number of microcosms; and many of us were influenced by this model. Large issues were best studied in small settings.

The interest in the little community also came to have a political edge after 1968. The growing social critique of Swedish society focused on the alienation and anonymity of large-scale urban settings, as well as the bureaucratization of everyday life. For the counter-culture movement, small was beautiful, and the search for local community life became a search for cultural alternatives: small scale, dense and informal cultural settings. This utopia of togetherness fitted very nicely with the interactionist theories used by most of us. Cultural integration was created through face-to-face interaction. This was the kind of social stuff that created 'good cultures', rich in shared experiences, everyday rituals and habits. The search for good cultural models was also a way of empowering settings, which seemed marginal to the general developments in society. Rural villages, fishing communities and traditional working class neighbourhoods thus became models for social change. There was a strong emancipating element in the search for the little community. Ethnologists became champions of 'the little people' and 'the little community'.

This interest for the local was to dominate ethnological research during the 1960s and part of the 1970s, but by the end of the 1970s it had lost its leading position to an interest in identities and subcultures. The search for subcultures grew out of a wish to break down stereotypes of Sweden as a homogeneous society (or local communities as well integrated). The new concept was used to capture other social units and cultural systems than the local study, but here too the result was that some groups and milieu were considered 'more sub-cultural' than others: teenagers, children, women, workers, and immigrants. (Middle aged, mainstream, middle class men were consequently the least sub-cultural category that could be imagined.) There was still an aura of 'small is beautiful' in this tradition.

---

2 On closer examination, these coastal communities revealed a different reality.

The study of subculture began in an interactionist tradition but went on to follow a semiotic path: from roles and scenes to codes and messages; it began to focus more on the expressive aspects of culture: style, taste, codes, identity markers, and the like. A central concept in the study of sub-cultural identities and boundaries was the concept of 'culture-building', that is the analysis of how different groups continually constructed and transformed a collective image and lifestyle. The Marxist influences, mainly in the form of cultural Marxism developed by British scholars like Raymond Williams, E.P. Thompson and their pupils in the so called 'Birmingham school', also created a new interest in processes of cultural domination and subordination. These Marxist influences were often rather eclectically blended with ideas of hegemony taken from such different scholars as Antonio Gramsci, Norbert Elias, Pierre Bourdieu and Michel Foucault. The linking of class and sub-cultural studies mainly took the form of two rather different genres: the study of bourgeois culture as a hegemonic process and 'the making of Swedish working class cultures' within the Thompsonian tradition.

There were striking differences in the way these studies were framed and delineated. Working class culture was mainly studied in the form of community studies, whereas bourgeois culture was analysed through a bricolage of materials at a national level. Another effect of this research strategy was that working class culture much more often was studied through oral history, whereas bourgeois culture was analysed through memoirs, etiquette books, diaries, mass media material, creating a bricolage approach. Just as the study of peasant culture had previously drifted into a devolutionary search for 'a golden age' of classic forms, working class studies tended to focus on the heroic age of early class formation often seen as a 'purer' form of class culture than, for example, the periods after the Second World War.

Experts on the Everyday

The new topics and perspectives of Swedish ethnology in the 1960s and 1970s coincided with the rapid growth of universities, as the baby-boomers came of age. The number of students and departments grew and the discipline was no longer small, but rather medium-sized. Ethnologists also attracted a lot of media attention and gradually came to be seen as the experts on everyday life, scholars who could make the mundane exotic. This had to do with a new division of labour in the humanities and social sciences. Sociology, a big and dominant discipline on the Swedish scene, was torn by fights between Marxist theorists and an old positivist tradition. Neither camp had any greater interest in a more qualitative, ethnographic approach to everyday life. Swedish social anthropology had its focus on fields outside Europe, historians still showed very little interest in popular culture. Ethnologists carried on the old tradition of exploring the common folk and colonized contemporary Swedish everyday with very little competition.

In using the metaphors of 'discovering' and 'exploring' everyday life as a hidden but exotic culture, ethnologists drew on an ethnographic approach. It was in the habits, rituals, routines and traditions of daily life that 'culture' was embedded, as invisible to the actors 'as water to the fish', to quote a popular metaphor from the 1970s.

As European ethnologists in the 1960s oriented their research towards modern industrial society rather than the vanishing peasant traditions, the basic concept of folk culture was exchanged for the study of the everyday and 'the ordinary people'. These new concepts not only signalled a break with earlier generations, but also worked as a boundary marker against neighbouring disciplines. Ethnologists studied everyday life with a qualitative perspective unlike sociologists and historians. The concept of everyday life was used to show that the ethnological interest in society was not concerned with formal institutions and macro-structures. In reality the concept was not about the contrast between everyday and non-everyday phenomena, but a way of stressing a research interest focusing on neglected groups, activities, and spheres.

The interest in everyday life was also present in the emerging interdisciplinary field of British cultural studies. As in European ethnology, this interest in everyday life during the 1980s and 1990s was closely linked to the new concept of cultural creativity. Just like the notion of everyday life, the concept of creativity functioned as a counter-argument in cultural research. Groups and settings which at first glance seemed characterized by passivity, simply reproducing dominant patterns, were shown in the new studies to be active culture-builders and bricoleurs: busy reworking, elaborating, thickening their everyday lives with new meanings and routines. In both ethnology and cultural studies the notion of everyday creativity becomes not only a counter-argument against the presupposition that creativity is an elite resource, a cultural capital of the gifted, of artists, writers, intellectuals etc., but also against the notion that modern mass consumption is a passivizing, and homogenizing force in people's lives.

When consumption in the 1980s was redefined as 'cultural production' or 'symbolic production', the creativity of everyday life came to occupy a central position in this production process. The emphasis on cultural creativity as resistance must be seen as a reaction against an earlier discourse on the seduction of mindless consumers, that is it was locked up in the iron cage held together by market forces. Creativity became, in some ways, the weapon of the weak – a positive strategy of resistance. At times, the concept was used to describe strategies of dominance and oppression, while other terms were also employed for the creative and imaginative ways of control and manipulation.

The pairing of cultural creativity and everyday life can thus be seen as an attempt to redress the balance, and develop a more actor-oriented approach in studies of mass consumption. In the process, however, the 'when, where, how, why and who' of cultural creativity in everyday life was narrowed down, and produced a tendency to over-research some aspects and arenas and overlook others (see Löfgren 1997: 98 ff.).

Cultural creativity became a very positive concept. It was seen as enriching, elaborating, and 'thickening' everyday life, a process through which people make their daily existence colourful, unique, specific, distinct and above all positive. The perspective was usually reserved for those defined as the underdogs of the modern world: consumers, workers, women, teenagers, immigrants and other minorities, and this is where the concept is closely linked to ideas of counter-hegemony, to the tactics of resistance in the world of mass consumption or in postcolonial processes of globalization.

Some activities and people thus came to be seen as more 'everyday' than others. Another danger of this perspective was that everyday life was too easily seen as something 'down there' in a way that it seldom reflected on the problem of what the opposite of the everyday was supposed to be. It was forgotten that everyday life existed not only in retirement homes or on the streets but also in the corridors of power, and that there were everyday practices not only in immigrant suburbs but also in scholarly research projects. Researchers who said that they wanted to 'get out into everyday reality' missed the point that what is interesting about the everyday is that it is ever-present.

An Idyllic World?

The focus on everyday life and cultural creativity was not without risks. Ethnology got the reputation of being a fun discipline, loved by the media for reflecting upon 'what Swedes are really like'. This became problematic when the question of national identity returned in Swedish society, as a result of the influx of immigrants and subsequently refugees during the 1970s and 1980s. For decades 'the national' had not been an issue in Sweden as was regarded as a problem of the past; but now it became once again a contested terrain in identity politics. The fact that the Swedish national self-understanding has been highly a-historical and apolitical, but also rather idyllic, must be noted here. It is important to look at the ways in which different nations choose to narrate their history: there are a number of genres here. The making of the Swedish (and Nordic) welfare state is usually told as a light-hearted success story: the nationalization of modernity without wars and great class conflicts. To a great extent Swedish ethnologists have embraced the basic credo of modernity: life can always be improved and we should keep an optimistic attitude about the future, culture building is a creative and positive process. This was an optimistic, but also a naive genre, that many of us contributed to, as some inside and outside critics started to point out in the 1990s.

In the early stages of becoming an immigrant society, Swedish cultural practices were very much hidden in ideas of common-sense rationality and modernity: 'Those immigrants have colourful and exotic cultures, we represent an everyday normality'. A new argument emerged especially in the growing public sector, which handled relations with immigrants, from social workers to schoolteachers. In order to understand their culture we must learn to understand ours, first, or at least make it visible. The result was a renewed interest in 'discovering and deciphering Swedish culture'; a new market emerged with university courses and seminars on 'Swedishness' and inter-cultural communication. In the public debate and in the mass media Swedishness was explored from all angles, at first within the rather ironic framework of the inverted patriotism of the 1960s, and later as a process of ethnification. A classic journalist question to ethnologists in the 1980s would be: 'give me five examples of something typically Swedish'. Most ethnologists tried to resist the temptation to generalize; instead they focused on deconstructing the concept of Swedishness, and yet were drawn into the debate. In the media Swedish culture was reified into lists of favourite culture traits or a certain 'Swedish mentality'. The ironic distance, as for example in the frequent use of 'Swedish' in book and paper titles, turned out to be naive and it could be used as a mechanism of exclusion.

The new kinds of poststructuralist discourse analysis that came to dominate much ethnology in the 1990s made research more conscious of such power dimensions. Discourse analysis also brought back a historical perspective in many cases, and also brought questions of gender, racism and postcolonial issues to the foreground.

An Academic Mentality

The last century of ethnological studies saw the theoretical perspectives of evolution and diffusion developed and slowly turned into studies of innovation. Functionalism was the cornerstone for early community studies in the 1950s and 1960s, later replaced with a more interactionist approach (Marxism and structuralism had their impact during the 1970s), while the 1980s saw the development of a more eclectic kind of cultural analysis.

The typical ethnological dissertation of the 1950s was a study of the diffusion of a cultural element in the dimensions of time, space and social setting – a folktale, a ritual, or an agricultural tool. A dissertation from the 1970s was often a community study with a focus on social life, while a 1990s study could focus on identity formation or lifestyle in a group, a minority or a subculture.

How much continuity is there between the folk-life researchers of the early twentieth century and the ethnologists of the early twenty-first? What characteristics have an enduring power or are more ephemeral, what is shared by many or few, what is central or peripheral to the discipline? In ethnology as in other academic fields there is the recurrent ritual of celebrating ancestors, pointing to continuity and the value of a long history. Here is a Swedish discipline that entered the university system before sociology, psychology or anthropology. This could just be shallow rhetoric, what if European ethnology has undergone such major transformations during its century in Academia that it no longer is the same discipline as when it was formed? Does the common label hide profound differences? In many ways the radical transformations are striking, when it comes to choices of theory, empirical fields and analytical themes, but there are also some notable examples of continuities, although they are found mostly at the level of meta-theory and research practices, in the approaches and ways in which a topic is handled. Let me exemplify four of them: the interest in the mundane, the ethnographic approach, a combination of the historical and contemporary perspectives, and, finally, the use of a mobile searchlight approach.

'Do not neglect a single tiny thing', was Lithberg's admonition in 1918 and, as I have shown, the interest in small worlds and tiny details has survived in changing forms. In the academic division of labour ethnology it emerged as the discipline of the overlooked or ignored, the seemingly trivial – all that the historians found too mundane, the aesthetic expressions art historians defined as too popular, the narratives that did not qualify as literature and the everyday routines that did not interest the sociologists. For earlier generations of ethnologists the prefix 'folk', as in folk art, folk narrative, folk tradition, signalled this field, but when this prefix became too problematic during the 1960s it had to be replaced by other disciplinary markers. Everyday culture was one of them, as seen, and this world was populated by 'ordinary people': such concepts were used to signal to the world that ethnological interest in society was something other than that of macro sociology or the history of institutions.

The second trait has to do with the emphasis on ethnography and a fieldwork mentality. I remember my surprise when I came from history to ethnology and was told that as a student you were supposed to create your own research material, through interviews, observations and a search for sources that often were found outside the well-known historical archives I had learned about as a history student. This mentality was born out of folk-life studies as a collecting activity. The Lund professor Sigfrid Svensson once wrote that it was this chance of personal engagement that made him choose ethnology, research was something carried out in the field, and theorizing was always intertwined with ethnographic work. Even as a student you had an independent role in some sense. I would argue that this mentality has survived even in research not based upon active fieldwork. 'Field-working in the archive or on the Net' is a metaphor supposed to capture this explorative approach in which a wide range of source materials is often combined. It is this hunter-gatherer mentality that still characterizes much of ethnological research.

Swedish ethnology is also a historical discipline, it is often said. Fifty years ago, this was something that was so self-evident it didn't have to be mentioned, but the amount of historical studies has varied a lot during the last few decades. Today, most of the research deals with contemporary arenas, but that it is the possibility of combining a historical and contemporary perspective that characterizes the discipline. To apply a historical perspective is often an analytical choice; history is used as a tool for problematizing the present. To be able to choose or not to choose a historical dimension in a study has promoted a certain kind of academic reflexivity. What are the analytical potentials or drawbacks of such a dimension in a given project?

Historical analysis in ethnology is often used in the work of deconstruct notions and phenomena that are taken for granted in contemporary society or else to problematize nostalgic ideas about 'past and present'. The historical perspective is also used to discuss problems of cultural translations between different eras and the dangers of academic anachronisms. Can a concept like identity be meaningful in the study of seventeenth century society? To what extent do concepts of home, work or leisure change meaning through different epochs?

When ethnologists talk of a historical dimension in their discipline they may mean different things. First of all, there is a need for a historical and ethnological knowledge of 'the past'. How far this past goes can be debated. Fewer and fewer ethnologists work with long historical perspectives, but as noted, it is not simply a question of how much cultural history should be included in ethnology but also what strategic advantages can be gained from a historical analytical dimension.

The ethnologist Börje Hanssen coined the expression 'the movable searchlight' as a basic ethnological tool. By this he meant the uses of a flexible set of perspectives, concepts and approaches that would scan the terrain for different phenomena and an awareness of how certain themes or problems would be shadowed by the more visible ones. He was interested in the often unconscious ways in which we come to overlook and neglect certain fields, materials or topics.

This calls for a constant and critical revision of ethnological work. Where is the searchlight concentrated right now and how do we blind ourselves through our research routines?

European ethnologists, just like other scholars, have always been busy developing concepts for change or cultural stasis. Looking back at the history of concept production in the study of culture, it is easy to see how trends develop and foci shift. There were the early days of developing all sorts of concepts for different forms of evolution and diffusion, there is the constant search for labels and metaphors for cultural mixes and confrontations, from acculturation and assimilation to biculturalism or cultural compartmentalization and then on to processes like creolization and glocalization. In the heyday of constructivism, cultural processes relied heavily on active verbs of crafting: building and constructing formations, but also on flux, flow and fragmentation.

Metaphors for cultural processes have been borrowed from very different fields like biology, linguistics, mechanics, and communication technology. Therefore, language is an important aspect when it comes to expressing concepts and perspectives, cultural phenomena. In other words, what happens when concepts are transplanted from one arena to another?

In 2004, a workshop on missing cultural processes was organized in Lund in order to reflect on such issues. Participants were asked to invent or rediscover cultural processes that could be used to focus on neglected types of cultural dynamics. The workshop (see *Ethnologia Europaea* 2005: 1–2) revealed a heavy emphasis upon processes of speedy renewal, fluidity and hybridity that has tended to skew scholarly attention towards what Elizabeth Shove has called 'a contemporary preoccupation with the explicit, the visible and the dramatic'; add to this a frequent use of theatrical metaphors, performance, setting, dramatization and choreography. Here we also find a focus on constant change emphasizing processes of cultural fragmentation, de-territorialization, and individualization. It was not denied that these processes were on going phenomena of contemporary society, but it was stressed that ethnologists lacked a large body of empirically bound research that could help understand the cultural forces of cohesion, stability and routinization that prevent daily life from spinning completely out of control.

There was a call for more attention to the types of cultural processes which turned the dramatic, exotic or explicit into inconspicuous elements of the mundane or were taken for granted. How do activities turn into routines, how are new cultural forms accommodated and coordinated with existing patterns? Sometimes, these processes are described in a derogatory way in terms of banalization and trivialization, but it is in the inconspicuous practices of daily life that small repetitive actions can work to subtly change larger social structures, and cultural values. This may confirm our love for small details, but as long as it is used as a back door entrance to larger social issues and does not trap us in a microcosm, it is acceptable to cultivate our ethnological obsession with smallness.

Great and Small Traditions

Like envy itself, 'discipline envy' is a mechanism or a structure. And it is this kind of structure, not the hierarchy of disciplines that endures. The prestige and power of individual disciplines vary over time; new disciplines develop; others fade away. On the contrary, envy, desire, or emulation, the fantasy of becoming that more complete other thing, is what repeats itself. (Garber 2001: 67)

Disciplines are more sacred in some sectors of Academia than others. In medicine and the natural sciences new fields are invented and old disciplines disbanded. In the humanities it has proved relatively easy to produce new disciplines, but almost impossible to dismantle old ones. This means that there is a set of disciplines emerging out of a nineteenth century order of things, but there are also new forms of interdisciplinary fields transgressing old boundaries, such as media research and cultural analysis.

Over the last century, Swedish ethnology has moved in different directions, borrowing frequently from some disciplines, while distancing itself from other. Anthropological theory was in great demand during the 1960s and 1970s, later on imports from cultural studies, sociology, and the new cultural geography became more striking. If we draw up an academic sociogram we can see how relations and foci change between generations of scholars. What kind of interdisciplinary dialogues are furthered and what are avoided? I remember once complaining of the ways in which historians often criticized ethnological research, when a colleague pointed out: 'Don't worry about their criticisms, it is when they stop criticizing us you should worry, by then we have become uninteresting.'

Small disciplines often define themselves as the eternal borrowers from the big disciplines that produce 'Grand Theory', but this is a simplified view of what happens when ideas travel between fields. Swedish ethnology has often been characterized by a rather eclectic combination of theoretical perspectives, conceptual frameworks, methods and techniques. Imports from other disciplines have been radically transformed by local practices, as for example in the kind of cultural analysis that came to dominate Swedish ethnology in the 1980s and 1990s and that in turn was exported to other traditions, such as sociology and history. Interestingly enough, the vague concept of 'the ethnological perspective' or 'the ethnological gaze' has been seen as the trademark of the discipline. (Ethnologists have also had a tendency to mystify this approach, talking about it rather in terms of 'a magic touch' that cannot be reduced to methodological recipes.)

In interdisciplinary projects ethnology is often called in to provide a different kind of analysis. The argument runs that 'an ethnological perspective would be helpful here', that is ethnology is seen as a certain style of doing research, a practice proper. During the last decade, ethnology has been seen as a versatile tradition, and it has cooperated with disciplines as different as medicine, genetics, political science, geography and economics. As a result, ethnology has been often incorporated into new multidisciplinary departments of studies in health, ethnicity or tourism. Much of the new labour market lies in such expanding milieu.

To conclude; is ethnology small, medium sized or large today? My argument is that although the discipline has expanded, it has retained a mentality of smallness that in some ways has been an asset. The fact that Swedish ethnology started out in a peripheral position, and its lack of a grand tradition of Founding Fathers has to some extent been an advantage, which has given the discipline great flexibility in reinventing itself a number of times.

Bibliography

Buttimer, A. (ed.) (1992) *History of Geographical Thought.* Special issue of *Geojournal.*

Ehn, B. and Löfgren, O. (1996) *Vardagslivets etnologi. Reflektioner kring en kulturvetenskap.* Stockholm: Natur and Kultur.

Frankenberg, R. (1966), *Communities in Britain.* Harmondsworth: Penguin.

Garber, M. (2001) *Academic Instincts.* Princeton: Princeton University Press.

Lithberg, N. (1918) 'Mortlar och pepparkrossare hos svensk allmoge', in *Fataburen,* 1–25.

Löfgren, O. (1997), 'Scenes from a Troubled Marriage: Swedish Ethnology and Material Culture Studies', *Journal of Material Culture,* 2(1), 95–114.

Lönnqvist, B. (2005), 'Paradigm and Field', *Acta Ethnographica Hungaria,* 50 (1–3), 339–47.

Chapter 7

The Hybridity of Minorities: A Case-Study of Sorb Cultural Research

Elka Tschernokoshewa

In April 2007 an article entitled '*Mehrsprachigkeit bringt das Gehirn auf Trab*' (Multilingualism gets your brain going) was published in the journal *Psychologie heute.* The article begins:

> For decades, bilingualism had a bad reputation. Even in the 1960s scientific opinion was that a second language over-taxed the brain. New studies show now that people who are bilingual have some advantages over the monolingual. People with only one native language demand too little of their brain – that is the result of an American research team led by Laura-Ann Petitto at the University of Dartmouth, who tested the language processing of mono- and bilingual test persons using the new method of near-infrared spectroscopy (NIRS). In principle, both people with only one native language as well as the bilingual use the same brain regions, mainly the speech centres of the left hemisphere, when they communicate in one of their languages. However, when bilinguals are asked to quickly switch back and forth between their two native languages, something surprising shows up on the NIRS-monitor: suddenly the speech centres in the right half of the brain – completely neglected by the monolingual – are increasingly utilised. This is the characteristic feature, in effect the signature of bilingualism, says Petitto. And something else astonished the scientists: bilingual test person's brain activity in the speech centres was always far above that of those monolingually raised.[1]

This article was published in the section 'Thema and Trends' and identifies something that we can observe as a trend in twenty-first century thinking. In the nineteenth century, the trend in relation to bilingualism or multilingualism was different. Here follows a quotation from the scholarly essay of the prominent Slavonic linguist August Schleicher from 1851. At the time, Schleicher taught in the department of Slavonic philology at the University of Prague; later, he assumed the renowned professorial chair in Jena. He was a very well known scholar and was regarded as a democrat. Still today, some of his texts belong to the standard literature of university Slavonic philology. In an essay entitled '*Über die Stellung der vergleichenden Sprachwissenschaft in mehrsprachigen Ländern*' (Concerning the status of comparative linguistics in multilingual countries) Schleicher concludes 'that even

1 Simone Einzmann, *Mehrsprachigkeit bringt das Gehirn auf Trab, Psychologie heute*, April 2007, S. 12.

in multilingual countries every inhabitant has always only one language which can have the holy value of a mother language: man should not become a hermaphrodite of nations; such creatures are, as are all hermaphrodites, monstrosities.' (Schleicher 1851: 15)

The phenomenon of bilingualism or multilingualism is directly connected with the situation of minorities, and members of minority groups such as Sorbs in Germany, Slovenians in Austria or Sinti/Roma (gypsies) in Bulgaria are, to a considerable extent, bilingual and multicultural. That is not only the case for these so-called 'old minorities' or 'resident minorities', but also for the 'new' or 'immigrant' minorities: Turkish immigrants in Germany, people from the former Yugoslavia in France, Indians in Great Britain. They have knowledge in their distinct languages as well as in the official languages of the nation states where they are living. Certainly, language knowledge varies considerably according to education level, occupational classification, and lifestyle. It also differs between the first, the second and the third generation. Nevertheless, a minimum of grappling with the language of the host country – be it in discussions with their own children who attend the public school, be it sitting alone in front of the television, or in an extreme case where one may discover 'that one understands nothing despite living here' belongs to the central experiences of migration.

For the Sorbs who live in Lusatia in East Germany, close to the borders of the Czech Republic and Poland, this concerns the Sorbian (a Slavonic language) and German languages. The people who speak Sorbian today also speak German. They move between both languages according to need (depending upon the circle of communication, or the topics of discussion; for pragmatic considerations, or just for fun). We observed how that functions in everyday life in an empirical study in Bautzen and the surrounding area, in the years 1998–2003. The results are published in *Auf der Suche nach hybriden Lebensgeschichten* (In Search of Hybrid Biographies) (Tschernokoshewa and Pahor 2005). Leoš Šatava's study *Sprachverhalten und ethnische Identität* (Language Behaviour and Ethnic Identity, 2005) examines how people deal with the two languages in depth. The Sorbian poet Róža Domašcyna (2006: 8) explains her language situation in the poem '*Windeierei*' (Wind Chimes):

> that I first learned spoke only father
> that I learned five-and-a-half days later spoke only
> mother
> I live rather in the third[2]

In this respect, in an article titled *Grenzgängerin zwischen zwei Sprachen* ('Bordercrosser between two languages', published in the *Sächsische Zeitung*) it is said that 'Róža Domašcyna does not write either German or Sorbian. She artfully mixes both languages. She loans something here, newly translates something there.

2 *die ich zuerst lernte sprach nur der Vater/ die ich fünfeinhalb Tage später lernte sprach nur/ die Mutter/ ich lebe vielmehr in der dritten.*

Checks the temperature of the languages, as she puts it, and selects the one which fits best'.[3]

For the Sorbs, bilingualism is not such a new and passing, not a temporary, phenomenon. This is true not only of the educated such as ministers or priests, teachers and poets, but also for the general public, as it is documented in one of the first ethnological studies. Wilibald von Schulenburg mentions this in the foreword to his book *Wendisches Volkstum in Sage, Brauch und Sitte* (Wendish Folklore in Sagas, Customs and Manners), written between 1880 and 1931. Significantly, he does not examine the problem of bilingualism. Here are two excerpts:

> There are two things on the minds of the young girls in the Spreewald, singing and dancing. The many Wendish folk songs, but also German, were sung in the spinning rooms and also while weeding and whatever else. This folk singing was cultivated.
>
> ...
>
> While, then and later, I did not find any interest in my collection whatsoever in the higher levels of Wendish society, I could look forward to a warm welcome in Schleife (area of Rotenburg, Silesia), not only from Mr Welan but also from the villagers. Many a night we spent in the parlour of the friendly hosts, sitting at the dining table, telling tales. There I was with my Round Table. They all spoke good German (von Schulenburg 1988: VII–VIII).

Yet, the doubling, or multiplicity even, do not concern the language only. Members of minority groups such as the Sorbs in Germany combine disparate elements in their everyday lives. They move between cultural expressions of the majority community and their own minority community, developing a multiple perspective. Our empirical research produces evidence for this repeatedly: people here grew up with the good and bad spirits of two or more cultures. Although they can tell a number of stories about disrespect and discrimination coming from the German side, they have friends in all social circles. Theirs is a very ambivalent life. Students at the *Sorbisches Gymnasium* (Sorbian high school) in Bautzen write essays about Jurij Brězan and Thomas Mann; play rock music with Sorbian, German and English lyrics, have friends in Prague (with whom they can communicate effectively in Sorbian), as well as in Leipzig (where many choose to go to study); they create their homepage in Sorbian, and have a virtual Sorbian village on the internet. The school choir sings a colourful mixture of Sorbian, Russian and Bulgarian songs. At the high school graduation ball they dance to modern, mostly English songs, as well as their Koło (quite like the way it is danced in Croatia and Serbia). They apply for entry to third-level education all over Germany, as well as abroad. From their places of study or training, they remain in contact using mobile phones and emails, and communicate in Sorbian. They meet in Bautzen at Easter or over Christmas, they take part in the traditional festivities, and organise their own alternative parties. Then they set out again. It is a life in between which is characterized by what can be termed transculturality.

3 Katharina Gräbner, *Grenzgängerin zwischen zwei Sprachen, Sächsische Zeitung*, 22.5.2007, S. 6.

If we understand culture as 'a programme of socially binding semantic interpretation of the reality model of a society' – to refer to the terminology of Siegfried J. Schmidt – members of minority groups such as the Sorbs in Germany have a double or multiple programme. As is well known, the aim of this programme is 'the reproduction of society and the symbolic control of individuals for the purpose of their communalisation' (Schmidt 2004: 85–100). In this sense we can say that members of a minority possess a multiple programme. These people uninterruptedly lead a multiple life. Multiple perspectivity – that is, the possibility to look at a problem from various points of view, and therefore to work on the problem with various 'cultural software programmes' – belongs to their everyday life. And this applies also to all those who do not command the Sorbian language, but nevertheless have certain experiences, sensibilities and competencies.

However, this same multiplicity has not always been able to fit in with the mainstream concepts of culture. For two centuries, German *Volkskunde* favoured unambiguousness and homogeneity of culture and ethnicity. As a discipline, it aimed to find and prove this unity and homogeneity, and it was connected with the intentions of the rising national modernity. In this vein, the fathers of German *Volkskunde*, as for example Friedrich Ludwig Jahn in his text *'Deutsches Volkstum'* (Characteristics of the German People), write: 'The purer a people, the better; the more mixed, the more like a rabble.' (Jahn 1813: 26) Accordingly, E. Hoffmann-Krayer formulated the foundations of the field. In his lecture of 1902 *'Die Volkskunde als Wissenschaft'* (*Volkskunde* as Science) he says:

> Tribal Volkskunde attempts to present the primitive views and folkloric traditions of a cohesive group, a community: be the group strictly or loosely defined, limited to a village, or a valley, or extending to a country, or an entire complex of peoples. What is important here is only the relatedness. (Hoffmann-Krayer 1902: 17)

Since the late 1960s, German *Volkskunde* has critically examined the term '*Volk*' repeatedly. This has resulted in the partial renaming of the field as 'Empirical Cultural Studies' or 'European Ethnology'.[4] The problems of bilingualism and multiculturalism are, however, still challenging for established ethnological scholarship. The idea of a homogenous or pure culture remains especially persistent in the public arena, in the media, and in politics.[5]

Sorbian Ethnology/Cultural Research as Island Research

In 1965 the well-known Sorbian ethnologist Paul Nedo published an essay with the title *'Sorbische Volkskunde als Inselforschung'* (Sorbian Ethnology as Island Research). The essay was published in the section 'Discussion' of the journal Lětopis and might have been intended as a provocation. In his essay, Nedo sketched the genesis and the criteria for a notion (today we would say a conceptualisation) of

4 Cf. Bausinger et al. 1978.

5 This is, for example, substantiated in a study of the German language press. Cf. Tschernokoshewa 2000, 2004.

Sorbian culture (or the ethnicity) as an island, and thereby also Sorbian *Volkskunde* as Island Research. At the beginning of his exposition he offers a brief definition: 'By an ethnic island we understand a small, more or less closed ethnic group, which as a rule lives for hundreds of years in the midst of another ethnic group.' (Nedo 1965: 98) Territorial isolation, permanence, stability, homogeneity are fundamental parameters of the figure of an ethnic island. What applies here as the most important criterion – as Nedo emphasises – is the own language of the island population. For that reason, such phenomena as folk songs and folk fairy tales play such a prominent role in the argumentation for the island. Further, for this conceptualisation, a homogenous village-farming population is assumed. Some institutions, such as the church, have a leading function as 'producers' of the island culture. Nedo emphasises also the conservative power of the island and also especially the ideological underlining and objective of island research. He writes:

> The ethnological concern with the island problematic in Germany increased considerably during the 1920s and 1930s, in connection with the intensified support for German settlements in East and South-East Europe by the authorities and institutions of the Reich. It was called 'deutsche Sprachinselforschung' (German language island research). This is also the title of the first attempt at a theoretical and methodological summary by Walter Kuhn. It is sufficiently known that all of these efforts for the so-called German folk groups were permeated by nationalistic and revanchist ideas, and had increasingly come under the influence of Nazi-imperialistic ideologies. (Nedo 1965: 99)

The concept of the ethnic island leads, as Nedo (1965: 100) says, 'to a disquietingly narrowed point of view, to completely one-sided selection of material and to distortions in the presentation, which amount to falsifications of reality'. Nedo also draws attention explicitly to the processes of industrialisation in the 19th and 20th centuries, to problems of bilingualism, similarities between Sorbs and West-Slavs, Sorbian-German relations, the socio-economic changes after 1945, as for example the construction of large-scale enterprises of the power-producing industry in Lusatia, and the inclusion of the Sorbs in this process, as well as the development of modern lifestyles.

Nedo's critical dispute with the conceptualisation of the Sorbian island was given little attention in the discussion during his lifetime, although he was one of the best-known Sorbian researchers. He was also an active scholar outside of Lusatia and directed the Institut für Ethnographie at Humboldt-Universität Berlin. Despite this, his text was not able to shift the paradigms, and the notion of the Sorbian island remained more or less unchanged until the 1990s. This is further supported in a famous essay by the Sorbian poet Kito Lorenc:

> If one is talking about Sorbs or the Sorbian, one also tends to approach this topic even today with the traditional picture of the 'Slavonic island in the German sea'. If I think about my path to – and in – Sorbian literature, this also was, for me, at first, the most conspicuous, but ultimately the most question-able [sic]. (Lorenc 1999: 409)

Lorenc critically examines the concept of the Sorbian island and the German sea and proposes the metaphor of the Wendic Seafaring (the essay bears the provocative title of 'Die Insel schluckt das Meer', 'The Island Swallows the Sea'). The same has authored also a play called *Wendische Schiffahrt* (Wendic Seafaring), which was premiered in 1994 at the Deutsch-Sorbisches *Volkstheater* in Bautzen. Speaking of this play, Walter Koschmal, the Regensburg Slavonic linguist notes that:

> In [Lorenc's] *Wendische Schiffahrt*, the Sorbian remains, the Sorbian culture remains, a substratum of the German language text. The decisive difference to the Old Slavonic comparative paradigms resides in that the biculturalism, which in its fundamental facts is not at all new, is here being poetically functionalised. For the first time, a poetic dimension is reclaimed from biculturalism. (Koschmal 1998: 89)

In cultural research, the examination of the concept of the Sorbian island has only recently gained importance (Tschernokoshewa 2000, Keller 2002). The conceptualisation of the Sorbian island is confined within the framework of the homogenising paradigms. To a certain extent it presupposes the figure of the German sea in which this island is found. Both – the island and the sea – have a substantial character when used as metaphors for culture. Additionally, they are parts of one and the same construction process. Accordingly, the only option that seems to be left for the island is that it will get smaller. What happens to the 'shrunken' parts, flooded by the sea, has not been thought of in the homogenising conceptualisation. Nor has it been asked whether something of the island will remain below the sea, or whether the sea itself has gained in substance by the flooded island. In other words, questions of mixings and solidarities have not been raised. The homogenising conceptualisation makes it more difficult, or hinders the possibility, to think about the relationship between different cultural expressions. It leads willingly or not to the notion of the 'sinking of the last remnants … in the great flood of Germanness' (Andree 1874: III). This prophecy appears in Richard Andree's *Wendische Wanderstudien. Zur Kunde der Lausitz und der Sorbenwenden* (Wendic Excursions: Studies on Lusatia and the Sorbian-Wendish), which was published in 1874.[6] Along the same lines, a diagram in Paul Nedo's essay shows the borders of the Sorbian language territory at the time of the reformation, after the Thirty-Years-War, in 1886, and in 1930. It is clear from the diagram that the island is shrinking. Perhaps journalists and intellectuals, who are encouraging 'reliable information' on 'how many people *still* command the Sorbian language' have this image of the shrinking island in their minds, as a recent article in *Der Spiegel* has suggested.[7] What is striking here is the use of the word 'still'.

Small Cultures and Minority Cultures

There is in fact a widespread notion that the Sorbian culture is a small culture. This notion owes to the metaphor of the always-shrinking Sorbian island as much as it

6 Andree was the publisher of the journal *Globus*, and the co-founder and long-term director of *Kartographische Anstalt von Velhagen und Klasing* in Leipzig, where he collaborated on a physical atlas of the German Reich in 1877, and the *Allgemeine Handatlas* was published in 1881.

7 Stefen Berg, '*Sachsen für Sorben*', *Der Spiegel* 42/2007, S. 50.

comes from linguistics, and especially from a definition of the Sorbian language as the smallest Slavonic or the smallest among West-Slavonic languages. This idea is then projected onto other areas such as, for instance, songs and speeches. Thus, one will easily hear about the 'smallest Slavonic people'; or come across songs about 'the smallest twig on the great Slavonic tree'. The notion dates back to eighteenth century National Socialism, to a time when there were differing concepts (Förster 2007) and the 'smallest Slavonic people' image was reactivated. This is confirmed by a brochure written Paul Nedo and entitled *Die Sorben in der DDR. Vom Leben des kleinsten slawischen Volkes* (The Sorbs in the GDR. About the life of the smallest Slavonic people), and published on the occasion of an exhibition *X Weltfestspiele der Jugend und Studenten* (10th World Festival of Youth and Students, Berlin July 1973–April 1974).

In 1988, the prominent ethnologist Wolfgang Jacobeit used the same image in his foreword to Wilibald von Schulenburg's *Wendisches Volkstum in Sage, Brauch und Sitte* (Wendish Folklore in Sagas, Customs and Manners):

> That there was a German ethnologist in this difficult time for all that was Sorbian, who stood on the side of the 'smallest Slavonic people' on German ground, should continue to serve as encouragement to study closely the life and work of this man (Jacobeit 1988: VII).

Though Jacobeit puts the phrase in quotation marks, he does not acknowledge it nor does he expand on it. This is very common practice in academic texts, press articles, leaflets, as all sorts of public statements today. Thus, for instance, the leaflet *The Sorbs in Germany*, put out by the *Stiftung für das sorbische Volk* in 1998, begins with the following sentence:

> Did you know that East Saxony, also known as Upper Lusatia, and South Brandenburg, or Lower Lusatia, are the home of a tiny Slavonic nation? (1998: 5)

This formulation has acquired various connotations. Alongside traditional concepts of Sorbian culture which invoke 'the national' and insist upon purity, and which assume the equivalence of language, culture and nation, there are new ideas which are expressions of a wish to carve out the relational, the situational and the dynamic. Within Slavonic linguistics there is presently an interesting discussion. For the new ideas I would direct attention to texts by Walter Koschmal (1998) or the article by Christian Prunitsch (2004) entitled *Zur Semiotik kleiner (slawischer) Kultur* (On the Semiotic of Small (Slavonic) Culture). In cultural research it is becoming increasingly difficult to work with a language-centred concept of culture; especially when we are interested in culture as social processes and everyday practices, this approach seems to be unproductive.

Here I am going to focus on two aspects: firstly, the relationship between language and culture. In this concept of culture, the boundaries of a culture are equated with the boundaries of language usage; language is regarded as 'the pillar' of culture. Equating Sorbian language and culture is a received notion, which to a great extent ignores the nonverbal aspects of culture or relegates them to a secondary place. Additionally, this model shuts out people who do not speak Sorbian, although they have an identity-forming connection to Sorbian culture. There are highly differentiated worlds of experience, not covered nor exhausted by language. I am thinking here, for example, of the community-building effect of music, or of all the feelings and thoughts that arise when looking at old family photos where one's own great-grandmother wears the traditional Sorbian costume.

The second aspect concerns power constellations. For the study of everyday culture and especially for the relationship problematic, reference to comparative minority research proves to be particularly productive. With the formation of the German state in the nineteenth century, due to the principles of the nation state, national culture and national literature, the distinctive Sorbian culture became a minority culture. This was a highly significant process which tended towards homogeneity, uniqueness, otherness and the exclusion of difference. Carl Schmitt, who strongly influenced constitutional law and the politics of nation building in Germany at the beginning of the twentieth century, wrote:

> Every genuine democracy is based on not only that equals are treated equally, but, with inevitable consistency, that the unequal are not treated equally. To democracy necessarily belongs, firstly, homogeneity, and secondly – if necessary – the exclusion or extermination of the heterogeneous (Schmitt 1923: 13f).

Friedrich Heckmann offers an illuminating description of the same process. Ethnicity, a universal phenomenon of human social formation, with nationalism, takes on a new and outstanding importance as principle of political and social organisation. Per force of the principle of nation and nation state – within the framework of nationalism as a political ideology, with its norm of establishing nation states which are as far as possible culturally homogeneous forms – heterogeneous groups, which live or migrate into the state's territory, are turned into ethnic minorities (Heckmann 1992: 65).

The process of the formation of minorities is complex. Heckmann (1992: 65) also points to this aspect: pressure to adjust and assimilate, or open enmity towards ethnic groups, often has the tendency to evoke or strengthen resistance and ethnic group solidarity, and in this way it contributes to the constitution of ethnic minorities. If we want to understand the German-Sorbian relationship, we should not neglect this aspect.

National modernity set out with the ideas of cultural purity and homogeneity, and social structures were substantially built according to this idea. Consequently, two strategies of dealing with cultural difference have developed in modern societies: assimilation and exclusion (Zygmunt Bauman 1997). Sorbs in Germany have experienced both strategies, as have other ethnic minorities. They have also experienced resistance, consolidation, and new solidarities. This cannot be described adequately with small *vs.* big, because it is about a 'special kind of smallness', about a difference within a state, that is, within a political structure.

The problem of dealing with difference – be it within a state, within the European Union, or in the context of our relation to the Near East – can be described as the basic problem of the human community at the moment. In times of global circulation of products, people, and symbols, this is something that affects all people and all regions of the world. As Arjun Appadurai has it: 'The central problem of today's global interactions is the tension between cultural homogenization and cultural heterogenization' (1994: 328). After 9/11, the proclamation of the 'axis of evil', the war in Iraq, but also the Eastern extension of the EU, this problem has taken on a new intensity. Comparative minority research can provide insight into the ways of dealing with difference, and possibly also make visible further strategies and visions for dealing with otherness. Bilingualism, multiculturalism, or the hybridity of minorities can be a promising vision of the future for the human community. In this sense, minority research is important, not only for minorities but for the social system as a whole. Therefore, I wish to appeal for the strengthening of comparative minority research, and in particular of research both the 'old' minorities such as the Sorbs, Sinti and Roma in Germany, and the 'new' minorities such as immigrants from Turkey and Russia. It remains very difficult to bundle these researches together as they are located in quite different departments and have separate academic traditions. It should be noted also, that it is no less difficult to establish minority research at the level of research institutions.

From Island Research to the Analysis of Relationships

In September 2005, the *Sorbisches Institut Bautzen/Budysin* in partnership with the *Institut für Soziologie der Freien Universität Berlin*, held a conference on the topic *Beziehungsgeschichten. Minderheiten – Mehrheiten in europäischer Perspektive* (Stories/Histories of Relationships: Minorities and Majorities in European Perspective). At the conference, renowned scholars from Germany, France, Austria, UK, Czech Republic, amongst others, took part (see proceedings, Tschernokoshewa and Gransow 2007). In the papers presented, but also in the lively discussions during the conference, attention was drawn to how little research is being conducted on these relationships. The analysis of multilingualism, double-mindedness, and transculturality stands as a challenge for cultural research. Thus, it becomes increasingly evident, that we are not able to understand the life of people – for example in Lusatia – if we do not place these relationships at the centre of our research. This calls for a special research perspective that may be called hybridology. It is a perspective which attempts to observe and analyse the multiple coding, as well as the complex weave of relationships. The term hybrid has been developed in recent years, especially in connection with post-colonial discourse. Of particular relevance is the work of Stuart Hall, Homi Bhabha, Zygmunt Baumann, Frantz Fanon, Edward Said, Ulf Hannerz, Arjun Appadurai, and Jan Nederveen Pieterse as well as texts by Salman Rushdie and V.S. Naipaul. Alongside the term hybridisation/hybridity, terms such as creolisation, syncretism, collage, bricolage, transculturality have also been used. These notions share a common background, that is the idea of reunion. Ultimately, however, the issue is a new theory and politics in relation to difference and diversity. As Hall notes in *Old and New Identities, Old and New Ethnicities*:

> The notion that identity in that sense could be told as two histories, one over here, one over there, never having spoken to one another, never having anything to do with one another, when translated from the psychoanalytic to the historical terrain, is simply not tenable any longer in an increasingly globalised world. It is just not tenable any longer. People

> like me who came to England in the 1950s have been there for centuries; symbolically, we have been there for centuries. I was coming home. I am the sugar at the bottom of the English cup of tea. I am the sweet tooth, the sugar plantations that rotted generations of English children's teeth. There are thousands of others beside me that are, you know, the cup of tea itself. Because they don't grow it in Lancashire, you know. Not a single tea plantation exists within the United Kingdom. This is the symbolization of English identity – I mean, what does anybody in the world know about an English person except that they can't get through the day without a cup of tea? Where does it come from? Ceylon – Sri Lanka, India. That is the outside history that is inside the history of the English. There is no English history without that other history. (Hall 1991: 48–49)

For a long time, the term hybridisation has had a difficult status in German-speaking regions. This is only now being overcome in the framework of minority and migration research. We can cite sociologist Ulrich Beck, who analyses the processes of global modernity:

> Thinking and doing research within the trap of the idea that societal worlds are separated and organised according to the principle of nation states, excludes everything that falls between these inner and outer categories of organisation. What is between these categories – the ambivalent, the mobile, the fleeting, the being simultaneously here and there – is only made accessible within a framework of migration research taking the approach of transnational social spaces. (Beck 1997: 53)

Only now is this research perspective gaining in importance as the following texts clearly demonstrate: *Globale Kultur in Deutschland – oder wie unterdrückte Frauen und Kriminelle die Hybridität retten* (Global culture in Germany – or how repressed women and criminals rescue hybridity) by Mark Terkessidis (1997); *Umgang mit Differenz. Die Migrationsgesellschaft im Kontext globaler Öffnungsprozesse* (Dealing with difference: The migration society in the context of global processes of opening) by Erol Yildiz (2007); or, *Ethnizität und Migration* (Ethnicity and Migration) by Kien Nghi Ha (1999); especially also the text *Lust auf Sorbischsein* (A passion for being Sorbian) by Konrad Köstlin (2003). In a chapter entitled *Der hybride Anthropologe und die Collage als Erfahrung und Ausdruck* (The hybrid anthropologist and collage as experience and expression, in *Anthropologisch reisen* (Anthropological travel, 2002) the anthropologist Ina-Maria Greverus writes that what in the first half of the twentieth century was regarded as the 'terror of anthropologists' has gained plausibility. The 'hybrid anthropologist' has made this into the preferred field of observation. Mixedness, in fact, affects not only the ethnic or the national but also other parameters of culture or identity such as gender, age, and occupation. Accordingly, Greverus writes:

> Hybridisation is no longer today primarily the consequence of colonial rape; rather, it is more strongly shaped by the weakening of the national-culturally and ethnically dominated, master-servant – particularly master-maid – establishing of boundaries in the fields of inter-gender, inter-personal, and inter-national encounters. In any case, the voluntary exchange of cultural discoveries and traditions is a question of access to the cultural (and, with this, social and economic) capital of education. (2002: 26)

Concerning cultural research in the area of the Sorbian minority in Germany, the perspective of hybridity is more appropriate than traditional concepts of pure spaces, isolated islands, and small cultures. We can describe the numerous old and new discriminations against Sorbs in Germany, but also the joy and wealth of 'being other' and 'being double'. We can also ask questions about discrepancies in cultural understanding, and then analyse who does something, and when, and why they do it. For at the present time, there are important struggles over interpretation and distribution going on in this field. From 1998 to 2003, in collaboration with students of the Universität Bremen, we conducted an extensive study in the area of the Sorbian minority in Germany and made instructive observations; here, an excerpt from the text *Schulzeit* (Schooldays):

> Ultimately, in the process of our research, we received the impression of a clear discrepancy between what is officially understood to be Sorbian culture – that which is emphasised within the framework of maintaining traditions, folklorisation, and keeping the language alive – and that which is actually entailed in the everyday life of people in this region. (Altkämper and Schatral 2005)

In order to develop this research perspective further, considerations of what has been found in relation to other minorities are extremely helpful. I would draw attention especially to the work of Judith Okely on the Roma and the Traveller communities:

> Instead of attempting to legitimise the Roma through recourse to an original, autonomous unity in a mythical past, we should appreciate the cultures of the Roma as complex and innovative forms, which can be copied by migrants. (…) The Roma have always been bricoleurs: they have taken things out of the systems in their environment and changed their meaning according to what suited them. They made some things their own, whereas they rejected other things. As I have shown elsewhere, the cultural productions of the Travellers, be it in painting, music, or in story telling, do not come into being through imitation, but rather by choice. And they have, reciprocally, given form to the surrounding dominant cultures. (Okely 2006)

The concept and the methodology of hybridology are currently being developed. Analyses of everyday life, but also of the political level, make that urgent. It is important to note that the term hybrid always focuses on a relationship; it is about relationships of cultural phenomena, which show differences that have taken place historically, socially, politically, etc. The processes of differentiation can be described as processes of social construction (here, the question of power is central). Such constructions include own-foreign, Sorbian-German, man-woman, young-old, good-evil, etc. The term hybrid requires no fixed elements but rather a combination of disparate elements which, depending on circumstances, can shift their position or change their relationship to each other. Therefore, this is not about absolute differences between own and foreign but about the permeability of borders; the 'third space', and the partial presence of the one in the other. The term hybridisation allows for both the construction and transgression of borders that can be studied with excellent prospects.

From the hybridising point of view, differences are seen, researched and taken seriously. Cultural differences are studied not as naturally grown givens, but rather as historically developed constellations, so that they become recognisable as products of human history. This perspective is highly sensitive to differences between cultures and within cultures: without making these into absolutes, without taking these to be inherent or unchangeable; and, what is especially important, without inexorably connecting these to every single action in the life of a person or group. Therefore, if someone 'sings in Sorbian' that does not mean that she/he does everything 'Sorbian'. It is the task of cultural research to ask when and why someone sings in Sorbian, for instance, and what experiences, sensibilities, and competencies are connected to this.

What is significant therefore is the explicit and open recognition of difference, and, at the same time, the attempt to bundle together otherness and sameness conceptually. We no longer and simply ask 'what is different for the Sorbs', but also 'what is the same' for them. And if we find similarities – which we are likely to – then we should not to forget the question of otherness, and ask again on the basis of the views newly acquired.

Minorities and Minority Research as Trendsetters

It is easier to work with the hybrid point of view in culturally mixed regions such as Lusatia. However, there is more to it than just that. It is revealed more and more clearly that knowledge which has been gained in the research of minorities plays a decisive role when we want to understand how it is that culture and identity function today, as hybridity is a phenomenon that affects all people and all regions of the world. Hybridity is, so to speak, the signature of culture in global modernity. Or to paraphrase Jan Nederveen Pieterse, 'Globalisation' is 'Hybridisation' (Pieterse 2004).

Life is globally structured already. In place of the old certainties, diverse insecurities, ambivalences, breaches, and mixtures are breaking out. Everything is increasingly connected. Our time is the time of 'and'. No one is 'only one', few are going to be able to remain in one place and one occupation; the normal biography is more and more turning into a self-created biography; flexibility and mobility are demanded from everyone. And the exchange with speakers of other languages, people with a different complexion, and of another religion, is becoming an everyday routine for everyone. Cultural difference is ubiquitous. The question is which concept and strategies are going to prevail. Three basic patterns are in competition: cultural differentialism ('The Clash of Civilisations'), cultural convergence ('McDonaldisation'), and cultural hybridisation. Each paradigm represents a different politics of coping with difference. Thus, while cultural differentialism translates into a policy of closure and apartheid, cultural convergence translates into the politics of assimilation, and cultural mixing refers to a politics of integration that does not need to give up differences but seeks a dialogue of differences (Pieterse 1996, Tschernokoshewa 2001).[8]

In this sense, minorities, with their fragile, mixed identities are true pioneers of hybridity. Once again, to quote Judith Okely: 'Precisely with regard to the

8 Maybe in 2008 the third paradigm gains in importance, when in the EU 'The Year of Intercultural Dialogue' is proclaimed (www.interculturaldialogue.eu).

connection of cultural coherence, identity, and hybridity, the Roma have held a pioneering role for centuries' (2006: 36). Though, perhaps it is for this reason that this research approach is slowly finding acceptance in general developments of theory, since there are also struggles over power and resource distribution in the formation and development of theory. Things may change now as they have in the past with the women's movement and gender research, until these topics and research perspectives were academically established. And, after that, it is still a long journey till the knowledge of research makes an impact in the public sphere and in politics, and also has an influence on social structures. For a new post-national Europe it is urgent to think about the possibilities of allowing for and coping with differences within the community. At this interface, minority research can contribute towards the permeation of a new trend. Let us call this new trend: "Sharing Diversity".

Acknowledgement

This chapter was translated by Richard N. Myers and Cordula Karich.

Bibliography

Allkämper, U. and Schatral, S. (2005) 'Schulzeit: Jugendliche einer zehnten Klasse des Sorbischen Gymnasiums in Bautzen' in E. Tschernokoshewa and M.J. Pahor HG., *Auf der Suche nach hybriden Lebensgeschichten. Theorie – Feldforschung – Praxis.* Münster, New York, München, Berlin: Waxmann, 147–169.

Andree, R. (1874) *Wendische Wanderstudien. Zur Kunde der Lausitz und der Sorbenwenden.* Stuttgart: Maier.

Appadurai, A. (1994) 'Disjuncture and Difference in the Global Culture Economy' in P. Williams and L. Chrisman (eds), *Colonial Discourse and Post Colonial Theory.* New York: Columbia University Press, pp. 324–39.

Baumann, Z. (1997) *Postmodernity and its Discontents.* London: Polity.

Beck, U. (1997) *Was ist Globalisierung?: Irrtümer des Globalismus, Antworten auf Globalisierung.* Frankfurt/Main: Suhrkamp.

Bhabha, H. (1994) *The Location of Culture.* London and New York: Routledge.

Domašcyna, R. (2006) *Stimmfaden: Gedichte.* Heidelberg: Das Wunderhorn.

Förster, F. (2007) *'Die, Wendenfrage' in der deutschen Ostforschung 1933–1945.* Bautzen: Domowina-Verlag.

Geertz, C. (1996) *Welt in Stücken: Kultur und Politik am Ende des 20. Jahrhunderts.* Wien: Passagen.

Greverus, I.-M. (2002) *Anthropologisch reisen.* Münster, Hamburg, London: Lit.

Hall, S. (1991) 'Old and New Identities, Old and New Ethnicities.' in A.D. King (ed.), *Culture, Globalisation and the World-System: Contemporary Conditions for the Representation of Identity.* Basingstoke: Macmillan Education.

Heckmann, F. (1992), *Ethnische Minderheiten, Volk und Nation: Soziologie interethnischer Beziehungen.* Stuttgart: Enke-Verlag.

Hoffmann-Krayer, E. (1902) *Die Volkskunde als Wissenschaft.* Zürich.

Jacobeit, W. (1988), 'Vorwort.' in Wilibald v. Schulenburg *Wendisches Volkstum in Sage, Brauch und Sitte.* Bautzen: Domowina-Verlag.

Jahn, F.L. (1813), *Deutsches Volkstum*, Leipzig: Hildesheim-New York 1980 (Nachdruck der Ausgabe Leipzig 1813).

Johler, R., Thiel, A., Schmid, J. and Treptow, R. (2007) *Europa und seine Fremden. Die Gestaltung kultureller Vielfalt als Herausforderung.* Bielefeld: Transcript Verlag.

Keller, I. (2002) 'Sorbische Volkskunde als Inselforschung? Überlegungen zu einem "alten" Thema' in M. Simon, M. Kania-Schütz and Sönke Löden Hrsg, *Volkskunde in Sachsen. Personen – Programme – Positionen.* Dresden: Thelem, pp. 291–300.

Koschmal, W. (1998) '"Wendische Schiffahrt" in deutschen Gewässer: die bikulturelle Poetik des Kito Lorenc in historisch-komparativer Sicht', *Lětopis*, 45, 85–96.

Köstlin, K. (2003) 'Lust auf Sorbischsein' in *Im Wettstreit der Werte: Sorbische Sprache, Kultur und Identität.* Bautzen: Domowina-Verlag, pp. 427–45.

Lorenc, K. (1999), 'Die Insel schluckt das Meer', in *Zeitschrift für Slavische Philologie*, 58(2), 409–22.

Nederveen Pieterse, J. (1995), *Globalization as Hybridization* in M. Featherstone, S. Lash and R. Robertson (eds), *Global Modernities.* London: Sage, pp. 45–68.

Nederveen Pieterse, J. (1996) 'Globalization and Culture: Three Paradigms', *Economic and Political Weekly*, 31(23), 1389–1393, in *Cultural Ecology in International Studies*, Yokohama, Meiji Gakuin University, Inst. for Intern Studies, 1996, pp. 64–76.

Nedo, P. (1965) 'Sorbische Volkskunde als Inselforschung', in *Lětopis*, 8, 98–115.

Nghi Ha, K. (1999) *Ethnizität und Migration.* Münster, New York, München, Berlin: Waxmann.

Okely, J. (2006) 'Kontinuität und Wandel in den Lebensverhältnissen und der Kultur der Roma, sinti und Kále' in R. Toivanen and M. Knecht Hrsg, *Europäische Roma – Roma in Europa*, Berliner Blätter, 39. Münster, Hamburg, London: Lit Verlag, pp. 25–41.

Prunitsch, Ch. (2004) 'Zur Semiotik kleiner (slavischer) Kultur', *Zeitschrift für Slavische Philologie*, 63(1), 181–211.

Šatava, L. (2005), *Sprachverhalten und ethnische Identität.* Bautzen: Domowina-Verlag.

Schleicher, A. (1851), *Über die Stellung der vergleichenden Sprachwissenschaft in mehrsprachigen Ländern.* Prag.

Schmidt, S.J. (2004) 'Kultur als Programm – jenseits der Dichotomie von Realismus und Konstruktivismus', in F. Jaeger and J. Straub Hrsg, *Handbuch der Kulturwissenschaften. Paradigmen und Disziplinen, Bd. 2.* Stuttgart: Verlag J.B. Metzler, pp. 85–100.

Schmitt, C. (1923) *Die geistesgeschichtliche Lage des heutigen Parlamentarismus*, 8. Auflage 1996. Berlin: Duncker und Humblot.

Schulenburg von, W. (1931/1988), *Wendisches Volkstum in Sage, Brauch und Sitte.* Bautzen: Domowina-Verlag.

Terkessidis, M. (1997), 'Globale Kultur in Deutschland – oder Wie unterdrückte Frauen und Kriminelle die Hybridität retten' in A. Hepp and R. Winter Hrsg, *Kultur – Medien – Macht. Cultural Studies und Medienanalyse.* Opladen: Westdeutscher Verlag, pp. 237–52.

Tschernokoshewa, E. (2000a), *Das Reine und das Vermischte. Die deutschsprachige Presse über Andere und Anderssein am Beispiel der Sorben.* Münster, New York, München, Berlin: Waxmann.

—— (2000b) 'Minderheiten als Grundfigur der globalen Moderne: Warum, warum nicht?', in R. Alsheimer, A. Moosmüller and K. Roth Hgg, *Lokale Kulturen in einer globalisierenden Welt. Perspektiven auf interkulturelle Spannungsfelder.* Münster, New York, München, Berlin: Waxmann, pp. 223–39.

—— (2001) 'Fremde Frauen mit und ohne Tracht: Beobachtung von Differenz und Hybridität' in S. Hess and R. Lenz Hg, *Geschlecht und Globalisierung. Ein kulturwissenschaftlicher Streifzug durch transnationale Räume.* Königstein/ Taunus: Ulrike Helmer Verlag, pp. 56–77.

—— (2004) 'Constructing Pure and Hybrid Worlds: German Media and "Otherness"', in U. Kockel and M. Nic Craith eds, *Communicating Cultures*, Münster, Hamburg, London: LIT Verlag, pp. 222–42.

Tschernokoshewa, E. and Gransow, V. (2007) *Beziehungsgeschichten. Minderheiten – Mehrheiten in europäischer Perspektive.* Bautzen: Domowina-Verlag.

Tschernokoshewa, E. and Pahor, M.J. (2005) *Auf der Suche nach hybriden Lebensgeschichten, Theorie – Feldforschung – Praxis.* Münster, New York, München, Berlin: Waxmann.

Yildiz, E. (2007), 'Umgang mit Differenz. Die Migrationsgesellschaft im Kontext globaler Öffnungsprozesse', in E. Tschernokoshewa and V. Gransow (eds) *Beziehungsgeschichten. Minderheiten – Mehrheiten in europäischer Perspektive.* Bautzen: Domowina-Verlag, pp. 49–62.

Chapter 8

Turning the World Upside Down: Towards a European Ethnology in (and of) England

Ullrich Kockel

Writing about England in the late 1990s, John Widdowson, then director of the Centre for English Cultural Tradition and Language at the University of Sheffield, noted that 'as we approach the millennium the outlook for the survival, let alone the proper development of folklore studies in the higher education sector, is decidedly bleak' (Widdowson 1997: 181). This rather sombre assessment, coming almost a generation after his American colleague Richard Dorson discovered, when attending international folklore congresses, that 'the Continental folklorists regarded the state of their subject in England as a joke' (Dorson 1980: 7), highlights the persistently marginal status of that subject in this country. Scholars in other disciplines often tend to hold curiously antiquated prejudices about folklore – a fact that might itself in time become a topic for folkloristic inquiry. At international conferences in the third millennium, it remains notable that England is not only underrepresented (relative to other countries), but is represented mostly by non-English scholars. Of seven participants representing English institutions at the 7th SIEF congress in Budapest in 2001, two were from Germany, and one each from Ireland, Scotland and the USA.

The Republic of Ireland, Northern Ireland, Scotland and, more recently Wales, have taken a somewhat different route. While this essay makes some reference to their situation, as appropriate, developments there cannot be discussed in detail. Unlike in England, folklore in these areas has, over the past decade or so, aligned itself to some extent with European ethnology – a term still little understood in England. However, there is a ray of light that might ultimately dispel the gloomy prospects invoked by John Widdowson some years ago, and help to advance the agenda for the kind of change that he proposed in his essay (Widdowson 1997: 184ff.). In 2001, the Economic and Social Research Council (ESRC) for Great Britain awarded funding for a series of seminars to explore the state of European ethnology in the United Kingdom, and to establish the foundations for a developmental framework, including research programmes and postgraduate training. The present essay reflects on the origins, trajectory and current position of ethnology and folklore in England specifically. Following a brief excursus into the origins and meanings of 'ethnology', the first part sketches the development of folklore studies in England, its main genres, and its past and present institutional position. The second part considers European ethnology in Britain at the start of the new century. A look to the future, setting pointers for a European ethnology in which a (yet to be constructed) English ethnology has a role to play, concludes the essay.

Ethnology as a Holistic Study of Society

In the introduction to his *Study of Man* (1898), the anthropologist Alfred C. Haddon defined ethnology as 'divided into several branches, the four more important of which are: Sociology, Technology, Religion, and Linguistics' (quoted in Fenton 1990: 182). He went on to say that 'we start with physical geography and find ourselves drawn into statecraft and political economy' (quoted in Fenton 1990: 183). The Ethnological Survey of the United Kingdom, commissioned by the British Association in 1892, had a similarly comprehensive perspective, not unlike the approach of *Allgemeine Statistik* ('general statistics') that characterised early ethnological endeavours in Germany. However, this view of ethnology never quite took root in England, for various reasons. Widdowson (1997: 182) points to a 'peculiarly English penchant for anti-intellectualism, and especially for disparaging the study of the history and traditions of the culture – purely on the smug and imperialist grounds that both are acknowledged as age-old and intrinsically superior to those of other cultures'. This attitude leads English people to dismiss their culture and traditions as 'the province of the rural, the old and the uneducated'. Linda Dégh (1978: 34) observed that 'European ethnology ... originated in the political struggles of various minority population groups for ethnic recognition through certain culturally distinctive features'. Consequently, 'the ... ethnography of the newly founded nations catered to the reinforcement of ethnic identity and the rise of national pride through the definition, propagation, and perpetuation of ethnic values.' There was, therefore, a greater propensity on the European periphery and in newly formed nation states like Germany, to engage in comprehensive ethnological surveys with a forward political purpose, as well as to collect ethnographic evidence for its own sake, or to preserve it from being 'quite trampled out by the iron horse' ('Old Folk-Lorist', writing in 1876, quoted in Dorson 1980: 8). One can certainly take issue with the kind of purpose that motivated much of this earlier research activity, but the key point to note for further consideration is that even in historical perspective, the *raison d'être* of ethnological and folklore studies in Europe was never exclusively antiquarian, although '[t]he roots of ethnology tend to lie in the field of antiquarian learning' (Fenton 1990: 177). Whether and how well this antiquarian image may have suited other disciplines in competition for increasingly scarce public funding is another matter.

As the precursors of European ethnology evolved, different countries chose their individual labels for these new disciplines. 'Efforts to find a generally accepted name ... have been part of the growth of the subject in general' (Fenton 1993: 10). At a meeting of Scandinavian and Finnish scholars at Jyväskylä/Finland in 1969, there was agreement that *folkliv* should be replaced by *etnologi* as the official academic term for the subject. A year later, German scholars, although by no means unanimously, found *Europäische Ethnologie* a suitable term for a reconstructed *Volkskunde*. However, a decade earlier, when an Institute of Dialect and Folk Life Studies was established at the University of Leeds, the Institute's founders thought it unwise to 'use the word ethnology in English studies, as we probably wouldn't have got that through the Senate and other committees' (Sanderson 1970: 104, quoted in Fenton 1993: 10). In the 1990s, the term 'ethnology' came into use at some Scottish and Irish institutions, most recently also in Wales. Whether this signals a substantive change of direction, as it did in Germany and the Nordic countries, or merely a pouring of old wine into new bottles cannot be discussed here. However, it might be fair to say that, as yet, we can see a bit of both happening, which is a good sign. One of the chief advocates of an ethnological perspective in and on Scotland, Alexander Fenton, stated some fifteen years ago in a lecture to the Folklore Society that he saw 'folklore as part of the wider field of ethnology as the term is now construed by agreement between European scholars' (1993: 11). What about England, then?

Folklore Studies in England: Evolution and Genres

William John Thoms coined the term 'folklore' in 1846 'as a substitute for 'popular antiquities'' (Dorson 1980: 7), proposing a research programme for the subject that required collection of 'an infinite number of facts scattered in thousands of memories, till the harvest could be analysed and presented in the manner of Grimm's Deutsche Mythologie' (Fenton 1990: 179). He also emphasised the need for comparative study. A theory of folklore, put forward by Andrew Lang in 1873, proposed a shift in focus, from philology towards ethnology (Dorson 1980: 8). This approach revolved around the key concept of 'survivals', formulated by Edward Tylor in 1871, according to which 'the peasantry, lagging behind the educated classes, still clung to vestigial practices and ideas – survivals – once central in the life of prehistoric man' (Dorson 1980: 8). In 1878, Thoms, Lang and others who 'subscribed to the anthropological theory of folklore-as-survivals' (Fenton 1990: 179) founded the Folk Lore Society in London. The 'survivals' paradigm was challenged by the emergence of diffusionist theory, which explained cultural similarities between 'widely scattered peoples not by comparable stages of cultural evolution but by the slow creep of transmission from one culture to another' (Dorson 1980: 12). At the International Folklore Congress in London in 1891, the controversy gathered momentum, and about the time of the First World War, the evolutionist folklorists had not only lost their debate with diffusionism, but the cause and standing of folklore itself had gone into decline.

In the broader context of Britain and Ireland, Fenton (1990: 181f.) identifies two phases in the early ethnology of these islands – a first, common one, ending roughly with the First World War, and a second, decentralised one, where the different parts of Britain and Ireland developed along somewhat separate routes. Both phases have several strands, including antiquarianism and folklore, but also philology and etymology, historical geography, and, especially in the second phase, material culture. Emphasis on material culture increased as the collections for folk and open-air museums were initiated from the 1930s onwards. However, English folklorists were accused of having failed to examine 'the form, structure, function and context of their data' (Widdowson 1980: 443). At the same time, 'an unquestioning acceptance of the primacy of print ... carries with it a prejudice against the validity of the spoken word – the medium through which the majority of folklore is transmitted'

(Widdowson 1997: 182). This attitude is so ingrained in our students that one of the major tasks of the tutor is to convince novices of the value of gathering data outside the library and the internet, by talking to and observing people, or walking around a locality with open senses.

Widdowson identifies various genres of folklore. The first and, in many ways, foremost is the study of 'language and other modes of communication such as gesture and the full range of paralinguistic, kinesic and proxemic features'. Although, as Widdowson (1980: 445) observes, 'language is indispensable to the maintenance of tradition ... it has received little serious attention from English folklorists'. It should be noted, however, that in the two instances where folklore actually gained a foothold in English universities, Leeds and Sheffield, the initiative of linguists was crucial. The second genre of folklore is custom and belief (Widdowson 1980: 447). Folk narrative (448), music, dance and drama (449), and material culture (450) complete the canon. These are what in German would be called *Sachgebiete*, literally: factual areas, or the subject matter of ethnology and folkloristics. They describe what the subject is about. How it goes about them is a different matter.

'The concept of ethnology is changing', as Fenton (1990: 184) observed well over a decade ago. A 'historical-descriptive' phase has given us collections and archives that are 'copious and representative of at least certain aspects of ethnological data'. In Britain and Ireland, this phase was linked with 'and partly motivated by the seeking of identity of the northern and western units of which these Islands are composed (as well as of some of the English provinces).' Next came a 'philosophical-speculative' phase (Fenton 1990: 185), in which the leading countries included Germany, Sweden, Hungary and the United States. The final decades of the twentieth century saw a third phase (Fenton 1990: 186), characterised, on the one hand, by a growing concern with issues of class and conservation, and, on the other hand, a 'coming home' of social anthropology from non-European locations. Several influences can be distinguished for this third phase. The first of these relates to the practice and purpose of collecting. SAMDOK (from Swedish *samtids dokumentation*, 'contemporary documentation') is an approach based on 'a sense of social responsibility ... that seeks to find the identity of modern communities and transmit it for the future, and so its orientation is more to the future than to the past' (Fenton 1990: 185). Covering the entire process from fieldwork to cataloguing, this is 'essentially an ethnological approach, as against the old-style museum approach which still often prevails in Britain, where documentation is no more than the paperwork that goes with a collected object' (Fenton 1990: 185). A second, closely related influence is the growing interest in social history since the 1970s. Researchers increasingly focus on questions of labour and women's history, urban life, or ethnic minorities. The renaming of museums in recent years is indicative of this influence. The Welsh Folk Museum became the Museum of Welsh Life, the Liverpool Labour History Museum became the Museum of Liverpool Life – in both instances, a broadening and refocusing of the institution's agenda was signalled by the new name. Fenton (1990: 186) identifies 'a third influence, reflecting contemporary anxiety about the environment and what man is doing to it', but does not explore this influence further.

Folklore Studies in England: Institutional Position

Dorson (1980: 12), writing shortly before the British government began to savage public services, including higher education, noted that 'the English universities, save for Leeds, have neglected folklore'. An 'Institute of Dialect and Folk Life Studies', established at Leeds University in 1960, operated there until 1984. Its collections are still held by the University, but the Institute became one of the early victims of Tory hostility to higher education. Unlike on the Continent or, to a lesser extent, in Ireland, Scotland and Wales, 'where struggling languages and cultures gave stimulus to folklore enterprises', English universities never established professorships and chairs in the subject. Dorson blames the eclipse of folklore in England on the conservatism of the universities. He cites John Holloway, who in 1976 argued that 'policy makers of the 1960s ... simply did not know ... of folklore as an academic discipline with possibilities for urban, ethnic, and working-class studies in contemporary society' (Dorson 1980: 12), lamenting the consequent 'loss of intellectual collaboration and so of intellectual vigour'.

There are departments with a core of regular full-time staff at universities in Dublin and Cork, Edinburgh and Aberdeen, and recently also at the University of Ulster. In England, however, with a much larger population base, there is currently only NATCECT, the National Centre for English Cultural Traditions at the University of Sheffield, with its significantly smaller base of mostly casual staff. This Centre, attached to the University's School of English, grew out of the Sheffield Survey of Language and Folklore, initiated in 1964, which contributed significantly to the Archives of Cultural Tradition, set up in 1968. Put together, the Survey and the Archives were in 1976 designated as a research unit under the name of 'Centre for English Cultural Tradition and Language', and this was changed in 1997 to the centre's present title. From the beginning under the leadership of John Widdowson, a distinguished linguist with a strong vision for developing folklore studies in England, the centre and its archives have become a resource of national and international importance. It is encouraging that, in these days of ever-tighter purse strings, the University of Sheffield, upon John Widdowson's retirement, had the vision to appoint another linguist, Joan Beal, at a senior level as full-time director of NATCECT. Even though the centre's staff base overall remains precarious, against the background of the past hundred years or so this bodes well enough for the future of the subject. The Centre now offers a range of undergraduate units, building up towards a half award, and a MA in Folklore and Cultural Tradition, comprising four modules that cover the entire canon of the subject, including an introduction to research methods.

A number of journals represent the field, serving different readerships. The oldest of these is *Folklore* (1878), the journal of the Folklore Society. *Gwerin*, associated with the Welsh Folk Museum, has been subsumed into *Folk Life* since 1963. Since 1957, the School of Scottish Studies in Edinburgh has published *Scottish Studies*, and the Review of Scottish Culture was launched in 1984. *Ulster Folklife* has been published in association with the Ulster Folk and Transport Museum since 1955, while the Dublin-based *Béaloideas* dates back to 1927. In England, the journal *Lore and Language* (1969) was launched in association with the Sheffield centre. Fenton (1993: 9) suggests that these journals are evidence of 'widespread, if not well coordinated, activity in ethnology in Britain.' In the same essay, he notes that in the late 1960s there were already 65 chairs of European ethnology on the Continent, while his own chair at Edinburgh, established in 1990, was only the second in these islands after University College Dublin. At the start of the twenty-first century, whether as folklore studies or as European ethnology, the subject remains without a designated chair in England (and Wales), compared with now two in Scotland, one in Northern Ireland, and a chair and several professorships in the Republic of Ireland.

European Ethnology in England: The Setting

In the 1990s, 'Englishness' experienced a discursive boom. John Widdowson (1997: 183) linked this phenomenon to the observation that the millennium was 'marked by cultural rites of passage throughout the world.' It may well be that a renewed interest in cultural traditions signals a certain coming of age of English society, although most commentators (e.g. Morley and Robins 2001) tend to interpret it rather as retarding into a kind of cultural siege mentality. I will come back to this point towards the end of the essay.

Long before the discourse of 'Englishness' began to proliferate, peripheral areas in Europe (and elsewhere) discovered culture and heritage as a resource for tourist development. Widdowson (1997: 184) notes that 'despite the strong public interest' in this regard, 'financial support for the core subject area is conspicuously lacking, both nationally and institutionally', except in the case of Museum Studies at the University of Leicester, a singularly successful programme living to a very large extent on that *Golden Milch Cow* of British higher education – 'full-fee'-paying overseas students. Apart from this, Widdowson is right in observing that the academic provision for the subject 'continues to be woefully inadequate in comparison with ... most other countries throughout the world.' Quite a number of participants in the new ESRC-sponsored network for European ethnology, mentioned in the introduction, have had to go to the USA or Canada to undertake doctoral research on England because of the dearth of facilities in this field at home. In the past, a small number of institutions have enrolled and supervised the occasional research student (like myself) in cognate departments, and continue to do so, and while this may have weakened the field by failing to create a canonically compliant cadre of dons, it may actually have given it added strength through scholars versed in the idiosyncrasies of neighbouring fields they can utilise in creative interdisciplinary confluence (more on this later).

Contrary to England, folklorists in the USA – England's primary peer nation, the Atlantic being perceived, in cultural terms, as a lesser gulf than the Channel – have long had a strong presence in the areas of public policy and public service. Since the 1970s, job opportunities in these areas have been systematically enhanced under the label of 'public folklore', which has a threefold connotation (cf. Welz 1997: 79):

1. 'Folklore in the public sector', that is, practiced outside higher education, but funded by the public purse and located in public institutions;

2. 'Folklore going public', that is, academic folklore playing a role in the public domain; and
3. 'Folklore as public service', that is, the lobbying for the recognition of folklore concerns, such as the appreciation of traditional cultural forms, as public tasks and duties.

More than two decades ago, Dorson (1980: 12) observed that 'a folklore-folklife-oral-history boom can be perceived now in the United States, responding to the interests in minority cultures and family trees epitomised by the success of Alex Haley's *Roots*. This has certainly been a key factor in raising the profile of folkloristics in the USA, and in the creation of significant employment opportunities for graduates in this field. Earlier, the emergence of the folklife studies movement, in virtual parallel with the new social history partly inspired by the French Annales school, had given an impetus towards a more holistic approach to folklore, which until then had been 'preoccupied with lore-manifested in the collection, classification and analysis of texts' while at the same time 'European scholars of *volkskunde* [sic], or regional ethnology, were more interested in the folk themselves' (Joyner 1999: 266). The reorientation meant in practice a growing emphasis on material culture in American folkloristics. This was replaced in the 1980s by a renewed focus on the text, but this time from a perspective that emphasised performance and communication rather than the formal aspects of the text. After a similar interlude of textual orientation, inspired by the encounter with cultural studies, European ethnology is now being encouraged to return to its focus on lived experience (e.g. Löfgren 2001) – that is, on the folk themselves. In the American case, Joyner (1999: 271) argues that for both the new social history and the new, text- and performance-oriented folkloristics, the 'fields they till have grown narrower and less fertile.' Moreover, the various 'excesses to which both paradigms are prone, under the ... influence of mathematics in one case and linguistics in the other, ... call for sober thought and stocktaking.' Joyner (270) accuses folklore scholars of having 'spent their energies in static synchronic analysis and become mired in microstudies of folklore performance in the ethnographic present', neglecting in the process 'the dimension of time – tradition'. He consequently calls for a 'reconsideration of the relationship between history and folklore', suggesting (271f.) that '[f]olklorists need history to help them understand the process of change in folk culture; social historians need folklore to help them understand the role of the folk in history, lest they degenerate into a gaggle of sycophantic yarn spinners.' Each field is acknowledged as having 'legitimate scholarly purposes of its own', but 'neither events nor structures, nor their interactions with one another, can be understood unless analysis is grounded in the attitudes and actions of real men and women, without whom there is neither history nor folklore' (272). Given that, historically, English folklore has had a strong association with linguistics (and philology, to reintroduce a term widely considered old-fashioned in English academia today), there may be a case for reconsidering here, too, the relationship between history and folklore. Again, more of that later.

European Ethnology in England: Fellow Travellers at the Crossroads

Linda Dégh (1978: 34) described what she called the 'folkloristic approach' in terms of 'the intensive, fieldwork-based study of a people's interrelated social, intellectual, and material aspects of culture in their ethnic context'. A similar, holistic emphasis on the field, on real-life evidence rather than primary reliance on textual interpretation and existentialistic introspection, has long characterised European ethnology, which is often somewhat flippantly portrayed as discipline 'between the chairs' of humanities and social sciences, with an almost post-disciplinary identity. At the same time, a fundamental crisis of this identity can been observed internationally (Johler 2001: 169, n. 16). It seems that European ethnology is constantly in need of reinventing itself, 'perhaps more so than traditional disciplines like history and sociology', but that its 'ability to adapt to changing social and cultural situations' may well be a reason for its relative success in recent years (Löfgren 2001: 151). In contrast with other fields, the distinction between basic research and more applied, practical and routine knowledge is not very clear-cut; at the same time, the high degree of engagement with praxis, for example through student projects, and the flexibility and versatility of its graduates are distinctive markers of subject identity (cf. Welz 1997: 81).

From a Scottish perspective, but considering Britain as a whole, Alexander Fenton (1990: 186) noted many years ago that it was 'as if anthropologists, ethnologically-orientated museum staff, the staff of ethnological institutes, and adherents of other related disciplines, were running along parallel roads with high walls between.' This 'fragmented approach to ethnology', he suggested, could be overcome 'if some magic means of integration could be found.' Can we pull down the walls of enclosure and work the ethnological commonage together again? In a later essay, Fenton (1993: 5) observes 'in ethnology an increasing degree of convergence with the methodology of the social sciences, especially social anthropology and sociology.' That he does not, at this point, mention cultural studies, with which some of European ethnology seemed to converge at the time, is perhaps not surprising. Cultural studies in Britain was then, and to a considerable extent still is, more oriented towards media and communication studies as a paradigm, continuing the textual emphasis noted above, while the vision Fenton set out in his lecture pointed more in the direction indicated by Linda Dégh in the quote above.

In the introduction to a recent collection of essays, Nigel Rapport reminds the reader that anthropological studies of contemporary Great Britain were 'originally the province of geographers ... and sociologists' (Rapport 2002: 3). Ireland was different in this regard, having long attracted the attention of North American anthropologists, and 'retaining its distinctive 'tropical' character' in the public imagination there. By contrast, any anthropological research in Britain was 'conducted in the shadow of a more proper anthropology elsewhere' (4). From this arose 'an uncertainty concerning the legitimacy (even the possibility) of undertaking anthropology in Britain' that has 'continued to dog the British institutional scene.' Although British anthropology has been 'coming home' for some time, this process was long presented partly in terms of simple expediency – cheaper, more accessible fieldwork locations – and partly in terms of ethical concerns over the anthropologists' right to interfere with culturally distant groups. Rapport's book, by contrast, 'makes Britain primary ... and does not see the need always to legitimise itself by drawing comparisons between Britain and 'other cultures'; nor does it feel that its lack of formal 'exoticism' requires special pleading' (2002: 5). The editor contends (7) that an anthropology in and of Britain:

> has the potential ... of providing some of the best that the discipline can offer because an anthropologist thoroughly at home in linguistic denotation, and familiar with behavioural forms, is more able to appreciate the connotative: to pick up on those niceties of interaction and ambivalences and ambiguities of exchange, where the most intricate (and interesting) aspects of sociocultural worlds are constructed, negotiated, contested and disseminated.

In concluding this collection of essays, the doyen of British anthropology 'at home', Anthony Cohen (2002: 326), suggesting that 'we should regard cultures as supposedly shared bodies of ideas, values, symbols and communicative systems within which people commonly misunderstand each other', indicates a possible convergence with a textually focused cultural studies perspective. However, he then points beyond that, to the fact that 'people live most of their lives in circumstances of particularity – family, friendship, work and collegial relationships – which themselves qualify those of generality, such as class, gender, ethnicity, nationality and so forth' (327). According to Cohen, anthropology's role within the orchestra of the 'humane sciences' derives from its 'peculiar competence ... to substantiate, inform and signal reservations about large-scale statements'. Not anthropological generalisations like those attempted by the early folklorists are the purpose of anthropological inquiry, but the description and analysis of specificity and difference – in other words, one may ask: ethnology?

In the already quoted lecture to the Folklore Society, Alexander Fenton referred to John Widdowson who, a few years earlier (1990), had 'deplored the fact that England had no major academic or public institution' that could 'function ... as an official forum for fostering a sense of regional identity and a proper pride in community and in maintenance of tradition' (Fenton 1993: 10). As 'Ireland, Scotland and now Wales have joined the cohort of European ethnologists', we are left with 'the problem of England' (11). Of course, this reference to 'tradition' came at the height of the debate over an 'invention of tradition', which, largely unintentionally, fed into a politically correct discourse decrying any appreciation of specificity and difference. As we are gradually beginning to recognise that this was also a hegemonial discourse driven by particular power political interests, we may, fifteen years on, be in a position to develop a more differentiated perspective on culture and tradition.

If folklore went into eclipse after the First World War, and anthropology developed as the study of 'exotic' cultures in far-flung places, Britain has not been entirely without an academically established field that concentrates on the study of culture 'at home', or rather, of modern, developed, industrial societies instead of backward, undeveloped, pre-industrial ones. Cultural studies as practiced in Britain has actually made a considerable impact on the way certain cultural expressions in societies like Britain are studied world-wide, even if the leading edge of this field appears to have moved across the Atlantic, and indeed further afield. Although a literary criticism approach to analysis still seems to dominate cultural studies in

Britain, and some institutes of European ethnology on the Continent may be attracted to this approach, there is also a more geographical and ethnological strand of cultural studies, developed, among others, by David Morley and Kevin Robins over the past two decades or so. In their introduction to a recent collection of essays on 'British Cultural Studies', they argue that the meaning of previously taken-for-granted terms like 'British' or 'English' is now increasingly blurred as the 'old hegemonies' have become 'clearly fractured' (Morley and Robins 2001: 4). Cultural studies needs to examine 'the internal tensions, divisions, and contradictions that have been held in check, but which increasingly seek to reassert themselves as the United Kingdom begins to break up' (7). Of particular concern in this context is 'the strength of the fearful 'Little Englandism' which ... continues to hold back any prospect of Britain's closer and positive involvement with Europe' (7f.).

While European ethnology has to compete for increasingly scarce public funding with these two cognate disciplines, anthropology and cultural studies, in particular, the concerns and research agendas expressed by, as it were, 'dissenting movements' within them – 'anthropology at home' and 'evidence-based cultural studies' – suggest much common ground, and a distinctive role for European ethnology. As Britain, and England within it, is becoming more evidently polycultural, a transformed – or, if you prefer: reinvented – folklore studies may well follow the path of its North American cousin folkloristics into the limelight of a 'public folklore'. With regard to studying 'interethnic contacts within multiethnic communities', Dégh (1978: 45) identified 'the ways and means of intercultural borrowing between interacting groups' as potentially 'one of the most fruitful fields in European ethnology.' If, in the process, we could overcome the 'peculiarly English penchant' (Widdowson 1997) for denigrating the academic study of one's own everyday culture, we might gain a thorough, evidence-based knowledge of 'England', and this in turn could help reduce that 'fearful 'Little Englandism'' Morley and Robins (2001: 7) talk about, allowing 'England' to take its place in the world with genuine and healthy self-esteem, rather than with a kind of post-imperial paranoia masquerading as moral superiority.

European Ethnology in England

Just like life itself, ethnology has its phases, and 'we must be aware of the expanding parameters of the subject in the present, as well as of its development in the past' (Fenton 1993: 7). In this essay, I have tried to sketch both, in broad outline, with reference to England. European ethnology is, as Fenton (11) quite rightly claimed, 'an evolving subject ... with absolute relevance to present-day conditions of life and with unrealised functional value in the deliberations of politicians and local authority planners and interpreters.' In the same lecture, he urged the Folklore Society to take, 'in close collaboration with a University Department ... in England, or even with more than one ... the bold step that would pull England into line with the rest of the British Isles and Ireland, and with the general European pattern.' Half a decade later, writing in the face of evidently unspectacular movement on this front, Widdowson (1997: 184ff.) offered 'a few modest yet basic suggestions which might form part of a strategy aimed at taking the subject forward in England':

1. 'Re-establishing folklore studies as an academic discipline' through a major, wide-ranging national campaign raising awareness of the practical applications of folklore in everyday life and work, and of key aspects of national and regional cultural identities.
2. 'Redefining the discipline' as one that recognises the cultural significance of a wide spectrum of cultural groups, including immigrant cultures, and of global forces and processes at work at the local level.
3. 'Making an inventory of existing resources' that would allow the best possible use to be made of collections already available.
4. 'Adopting a dynamic approach to the subject' that ought to give due credit to earlier approaches and material insofar as they contribute to contemporary studies now and in the future.
5. 'Establishing and developing new theory and methodology' as the foundation for a more holistic and dynamic approach.
6. 'Establishing and/or developing new fields for research and teaching', for example the role of folklore in politics or the sciences (see my earlier remark concerning academic prejudice against the subject itself), the folklore of war, or folklore in the contemporary media.
7. 'Exploring and clarifying the interrelationships of folklore and popular culture', including the danger that folklore 'may be swallowed up by the fashionable and burgeoning field of 'cultural studies', and may consequently lose its integrity, its individual identity, and its distinctive history and focus' on 'the myriad ways in which tradition monitors itself and operates within culture.'
8. 'Identifying sustainable funding mechanisms' which, given the location of the subject between the current funding councils for humanities and social science research, will require 'concerted efforts by the major institutions concerned'.
9. 'Maximising the potential of the subject' in the labour market by highlighting its profile as practically useful in the world of work; this can be done through courses aimed expressly at training a range of culture professionals able to apply their skills and knowledge across a variety of occupations.

Perhaps the late 1990s were a period of fermentation and maturation of ideas. In 2001, about a year before John Widdowson's 1997 article actually appeared in print, the ESRC granted funding for a series of research workshops, organised by a network of scholars empathising with the project of developing European ethnology in Britain. In its publicity material, the workshop series is described as aiming to:

1. Complement and enhance existing structures by bringing together scholars working in different fields within, or related to, European ethnology (e.g., social anthropology, cultural studies, folklore, geography, history, area studies);
2. Facilitate debate on theory and methodology, engaging researchers across these fields as well as research users; and,

3. Promote comparative research on historical and contemporary topics, placing the British Isles within a European context.

Objectives by which these can be achieved include the establishment of

1. A regular cycle of research seminars, organising international conferences and sessions at such conferences, and producing a regular newsletter available in electronic format;
2. A broad developmental framework for European ethnology in these islands, including the encouragement of interregional contacts, inter-institutional postgraduate programmes and, if viable, the production of a dedicated journal; and,
3. A database of European ethnologists working in the United Kingdom connected to a web site as a research resource.

A key emphasis in the workshops is on comparative perspectives in a European context, and on innovative approaches to evidence and social analysis. 'European ethnology' is defined as a discipline that 'investigates aspects of everyday culture, identity and historicity within their spatial, temporal and societal horizons', addressing particularly questions of 'cultural diversity and multiculturalism, social exclusion on cultural grounds, boundaries and territories, and the evaluation of policy effectiveness.' With this agenda, European ethnology is squarely referred back to its roots in *Allgemeine Statistik*, and this implies serious questions about the political involvement, role and accountability of the subject. 'European ethnology contributes to the development of methodologies for evidence-based multidimensional and complex social analysis, and has the capacity to inform policy and practice in a wide range of socio-cultural domains.' Being an interdisciplinary or, indeed, post-disciplinary discipline, it has a significant role to play in the formulation of epistemological foundations for rigorous interdisciplinarity that may, in time, replace the currently fashionable, free-wheeling dilettantism. At the grass-roots level, it is 'particularly suited to studying local-level interpretations and negotiations of global processes.' As with all political engagement, studying the dispossessed and disenfranchised and working out what makes them tick has two sides to it, depending on how the researcher may interpret terms like 'public interest' and 'public service' – and who exactly this 'public' is. In the past, work of this nature has frequently been undertaken and used, intentionally or otherwise, in ways that have brought the subject into dire disrepute. At the beginning of the twenty-first century, we need a very thorough understanding of where European ethnology has come from if we want to avoid history repeating itself. Therefore, a grounding in the history of ideas and their context, not just in theories and methods, must form an essential part of our undergraduate as well as postgraduate education.

European Ethnology of England

In the specific situation of England – to a lesser extent in other parts of the United Kingdom – we need to live, at least for some time yet, with the fact that the majority of European ethnologists and folklorists who have a formal qualification in the subject have earned their laurels abroad, while the vast majority of us are graduates of various more or less cognate fields who at some stage in their career have gravitated, for one reason or another, towards this one. With very few exceptions, we work in institutes or departments dedicated to a different disciplinary agenda, and as this is how we tend to butter our bread, we cannot completely ignore these different agendas. However, this is a situation typical of an emerging field or discipline, and in between bemoaning our predicament, we ought to find time to focus on the benefits of this situation, and what European ethnology in England might bring to the development of the discipline internationally. In this essay I have deliberately used 'field', 'subject' and 'discipline' interchangeably, as at this precise point in time it is difficult to discern which label is the most appropriate for European ethnology, not just here in England but also in the international arena. We are *en route* to somewhere – some national and regional ethnologies are further along the road than others, and European ethnology in England is only in its infancy. So is the anthropology of Britain (and Europe), and, albeit to a lesser degree, the branch of cultural studies that has rediscovered the more immediate everyday concerns of people 'on the ground'. What both of them seem to lack, by and large, is a sharp historical 'lens'. Fieldwork-based anthropology, with few exceptions, has not yet developed a sure grip on historical context, while post-Marxist cultural studies seems to have thrown out the historical baby with the materialist bath water. European ethnology and folklore have long had not merely an affinity, but a mutually enriching working relationship with social history in particular. The connection may have gone somewhat stale in places, but for a budding European ethnology in England the link with (regional and social) history would seem an obvious one to make.

This would also help European ethnology tackle a key sociological and political issue – the essentialisation of cultural traits that leads to the kind of fears expressed in 'Little Englandism' with a certain inevitability. European ethnology in England must not fall into the trap of providing rationalisations for such fears as its antecedents have often done in the past, here and elsewhere. Instead, what we need is the development of a culture-critical English ethnology, interdisciplinary, ethically aware, evidence-based, and policy-oriented, that provides a solid foundation for comparative studies – within England itself, with other regions of these islands, with elsewhere in Europe, and beyond. 'English' in this sense refers both to the cultural roots and contemporary expressions and refractions of 'Englishness', and to all the 'interethnic contacts within multiethnic communities' (Dégh 1978: 45) that make up English society at the start of the third millennium. To develop such an English ethnology, we must, as Widdowson (1980: 450f.) argued many years ago, 'first turn our attention to the collection of data in the fullest possible social context, and investigate its functions in, and relevance to, our daily lives.' This collection is not informed by romantic notions of disappearing worlds, but instead is forward looking and, as such, utopian in the best sense of that word. It is not about casting continuity in stone (or into the multi-media presentations of the heritage industry), but about recognising that the only continuous cultural factor is change. Tradition, then, is not about handing down the past in an unadulterated way but rather, about local control of the process of change (Kockel 2002: 192f.). In this sense, one can indeed agree with John Widdowson that 'each one of us is rooted in tradition', which 'informs our lives and infuses itself throughout our institutions' (Widdowson 1980: 451). And a European ethnology taking its grass-roots orientation seriously may, by focusing on tradition, very well help to turn the world upside down.

Bibliography

Cohen, A. (2002), 'The 'Best of British' – with More to Come …', in N. Rapport (ed.), *British Subjects: An Anthropology of Britain.* Oxford: Berg, pp. 323–29.

Dégh, L. (1978), 'The Study of Ethnicity in Modern European Ethnology', in F. Oinas (ed.), *Folklore, Nationalism, and Politics.* Columbus/Ohio: Slavica, pp. 33–49.

Dorson, Richard (1980), The Founders of British Folklore, in V. Newall (ed.), *Folklore Studies in the Twentieth Century: Proceedings of the Centenary Conference of the Folklore Society.* Woodbridge/Suffolk and Totowa/NJ: Brewer, pp. 7–13.

Fenton, A. (1990), 'Phases of Ethnology in Britain, with Special Reference to Scotland', in *Ethnologia Europaea*, XX, 177–88.

—— (1993), 'Folklore and Ethnology: Past, Present and Future in British Universities', *Folklore*, 104, 4–12.

Johler, R. (2001), 'Ach Europa! Zur Zukunft der Volkskunde', in G. König and G. Korff, G. (eds), *Volkskunde '00: Hochschulreform und Fachidentität.* Tübingen: Tübinger Vereinigung für Volkskunde, 165–80.

Joyner, C. (1999), *Shared Traditions: Southern History and Folk Culture.* Urbana and Chicago: University of Illinois Press.

Löfgren, O. (2001), 'Life after Postmodernity: Volkskunde in the New Economy', in G. König, and G. Korff (eds), *Volkskunde '00: Hochschulreform und Fachidentität.* Tübingen: Tübinger Vereinigung für Volkskunde, pp. 151–62.

Kockel, U. (2002), *Regional Culture and Economic Development: Explorations in European Ethnology*, Aldershot and Burlington/VT: Ashgate.

Morley, D. and Robins, K. (eds) (2001), *British Cultural Studies: Geography, Nationality, and Identity.* Oxford: Oxford University Press.

Rapport, N. (ed.) (2002), *British Subjects: An Anthropology of Britain.* Oxford: Berg.

—— (2002) '"Best of British!" An Introduction to the Anthropology of Britain', in N. Rapport (ed.), *British Subjects: An Anthropology of Britain.* Oxford: Berg, pp. 3–23.

Sanderson, S. (1970), 'The Academic Status of Folklore in Britain', *Journal of the Folklore Institute*, VII (2/3), 101–05.

Welz, G. (1997), 'Kontingenzen des Wissenstransfers', in C. Burckhardt-Seebass (ed.), *Zwischen den Stühlen fest im Sattel? Eine Diskussion um Zentrum, Perspektiven und Verbindungen des Faches Volkskunde.* Göttingen: Deutsche Gesellschaft für Volkskunde, pp. 79–85.

Widdowson, J. (1980), 'Folklore and Regional Identity', in V. Newall (ed.), *Folklore Studies in the Twentieth Century: Proceedings of the Centenary Conference of the Folklore Society*, Woodbridge and Totowa/NJ: D.S. Brewer, pp. 443–53.

—— (1990), 'English Language and Folklore: A National Resource', *Folklore*, 101, 209–20.

—— (1997), 'Backwards to the future? English folklore studies in the twenty-first century', in *Lore and Language*, 15 (1–2), 181–90 [published 2002].

Chapter 9

Ethnology in the North of Ireland

Anthony D. Buckley

My task is to survey some two centuries of ethnological inquiry in the north of Ireland. One problem is that, since the 1970s, droves of scholars have taken an interest in Northern Ireland, making it possible to cover this field only impressionistically. Another is that writers have traded under such different names as 'anthropology', 'ethnology', 'folklife', 'folklore', or even 'geography' or 'history' while studying essentially similar topics. I shall therefore risk using the term 'ethnology' to refer to the study of the everyday lives, actions and activities of the broad mass of the population.

Ethnology in Ulster has broadly taken four approaches, each of which has its value. One has been the simple gathering of empirical information. This approach predominated during much of the nineteenth century and persists to this day. A second has been the use of ethnological materials to idealise group identities, be these national, regional, ethnic or local. This celebratory approach blossomed during the Celtic Revival that began in the late nineteenth century, but it existed throughout the entire period. Another approach involved a quest for origins. Key notions here were that certain types of past culture might 'survive' into the present day, but also that they may 'diffuse' from one population to another. The fourth approach, which arrived rather suddenly in Ulster in 1972, saw the central task to be to discover the *meaning* that actions and objects might have in particular social contexts.

Early Empiricism

Much early ethnology in the north of Ireland consisted of the mere collection of information. In a recurring feature of Ulster ethnology, this information often consisted of bite-sized chunks – be they stories, tunes, brief accounts, finished narratives or even physical objects – that could be easily compiled, published or stored.

Such materials exist in the narratives of both travellers and residents (Carlyle 1849, Mr and Mrs Hall 1850, Plumptree 1817, Thackeray 1843). Many such people were interested to express sympathy or criticism of the poverty of Ireland; others, to represent Ireland's supposed archaic exoticism (Bell 1998, 100). A few such accounts are more systematic. Important here is Carleton's *Traits and stories of the Irish Peasantry* (1832). So are the copious memoirs on such topics as domestic life, material culture and folklore made by the soldiers, surveyors and scholars who created the maps of the Irish Ordnance Survey (Ordnance Survey Memoirs 1835–40, see Andrews 1975, Gailey 1982b).

The transcription of Irish music is also important. This really begins with the work of Edward Bunting. In the late eighteenth century, the custom of itinerant harpers visiting the great aristocratic houses fell into decline. In 1792, surviving harpers gathered in Belfast to perform their music. Bunting, transcribed their tunes (see Audley 2003, Harbison 1989, Maloney 2000, O'Sullivan with Ó Súilleabháin 1983), and these eventually provided the basis for a major genre of Irish song.

Newspapers often recorded ethnological information. Ronald Adams (1985) tells of the bonfires, cockfights and night burials recorded in early editions of the *Belfast Newsletter*. By the 1850s, the more scholarly *Ulster Journal of Archaeology* included similar information alongside its more central concern with archaeology and history. One typical volume of the journal (the seventh), for example, contained, 'The remarkable correspondence of Irish, Greek and Oriental legends' (O'Laverty 1859), 'On the heath-beer of the ancient Scandinavians' (Locke 1859), and (part of a series) 'Six hundred Gaelic proverbs collected in Ulster' (MacAdam 1859). Both the Belfast Natural History and Philosophical Society and the Belfast Naturalists' Field Club (founded respectively 1842 and 1863), whose prime focus was in natural science, also took an interest in ethnology.

By the twentieth century, this empirical spirit had become more sophisticated. In particular, T.G.F. Paterson, first curator of the Armagh County Museum (1935 to 1962) produced extensive archives and some still-useful publications relating to genealogy, local history, archaeology, folklore, place names studies, and dialect (see Abraham 2000, Evans 1971 (ed.) Evans 1975). Empirical research is, of course, the backbone of all ethnology, and the task specifically of *collection* has particular importance in such fields as narrative, music and material culture. While somewhat crude in the nineteenth century, this empiricism laid the foundation for subsequent, more theoretically orientated research.

Revival and Beyond

From the 1880s, Irish ethnology gained new impetus in the Celtic (or Gaelic) Revival, with its interest in recording and preserving narrative, music and language, all of which were conceived of as Ireland's *treasures*. A major force in Ulster was the Belfast lawyer and ardent nationalist Francis Joseph Bigger. Bigger became secretary and then president of the Belfast Naturalists' Field Club. In 1894, he revived the *Ulster Journal of Archaeology*, becoming its editor. He soon organized a multitude of events from folk ballad expositions to pageants, plays and processions, which he saw as integral to social 'regeneration' – a cause close to his heart. He restored Jordan's Castle at Ardglass, turning it into a centre for a peasant arts revival. (Dixon 1997, 40ff) Perhaps his most famous achievement was the *Feis na nGleann* (Festival of the Nine Glens of Antrim and Rathlin) in 1904 (Bell 1988, McBrinn 2002, 2004).

Bigger propagated traditional crafts, and encouraged them as commercial enterprises in the spirit of William Morris. For example, he helped set up toy-making workshops in Cushendall and Ballycastle (McBrinn 2002). The Ballycastle workshop included a mock-up of an old Irish kitchen with a 'turf fire, dresser covered in delft and pewter, its grandfather clock and settle bed.' At the shop opening, there sat an old woman spinning, and another grinding flour with a quern (McBrinn 2002, 45). Bigger himself, like the Revival generally, has been widely criticised for his 'fakelore' or 'folklorismus', the fact that many of its supposed traditions were *invented* (Gailey 1982c). Bell, however, generously claims there is nothing disreputable in merging old and new cultural forms, or vernacular and more elite traditions (Bell 1988, 1998).

In 1927, the Revival gave birth to the Irish Folklore Society followed by the Irish Folklore Institute (later the Irish Folklore Commission) under Séamus Ó Duilearga and located in University College Dublin (Briody 2005, 10). Despite an inevitable southerly orientation, these bodies maintained a continuing interest in Northern Ireland. For example, one of the Institute's first projects was to publish a collection of Tyrone folktales by Ó Tuathail (1933, see Briody 2005, 17–18). In the 1960s, the Commission employed Michael J Murphy to collect folklore in Northern Ireland (1964, 1965, 1973, 1975, 1976, 1989). And a recent project called 'Room to Rhyme' led by Séamas Ó Catháin and Críostóir Mac Cárthaigh of University College Dublin studied mumming as much in the north as in the south of Ireland (see Buckley *et al* 2007).

Estyn Evans and his Followers

While the Folklore Commission dominated the study of folklore in the Republic, ethnology in the North took a different turn, largely due to Estyn Evans and his students, among them George Thompson and Alan Gailey. Together, these three men were to create the Ulster Folk Museum.

Evans came to Ulster as an archaeologist (Evans 1935, 1966) though he quickly turned to ethnology, becoming Professor of Geography at Queen's University. He writes that, during fieldwork particularly in Rathlin and Donegal, he was impressed with 'the wealth of survivals in material culture and folk beliefs' (Evans 1988, 92). He brought to Ulster's ethnology a distinctive emphasis on the survival of ancient traditions and a focus on material culture.

'Survivalism' is a methodology still found in archaeology and in British and Irish folklore studies. It has two rather distinct origins. One is the romantic assumption, dating from the eighteenth century, that existing rural culture once *belonged* to a nation's ancient inhabitants. The other is 'social Darwinism', which formed the basis of most late nineteenth century social theory, which applied evolutionary thought to the study of human culture (see, for example, Tylor 1871).

One version of social Darwinism, espoused explicitly by Evans's teacher Herbert Fleure (1947) but found in Evans's own writings, was that certain types of object, custom or even physiological trait were found in the present because they had allowed a population to subsist in a particular geographical environment. These types of culture either 'survived' from a distant past or had 'diffused' from somewhere else. The ethnologist's task was to identify the origin of these types of culture, and to discover how they allowed people to adapt to a particular environment. In its emphasis on types (or species) of artefact and on the origin of these species, this kind of ethnology differs from mainstream historiography which emphasises the unique actions of individuals in particular historical moments. Much more to the point, however, Evans showed that, by describing and sketching material objects, and by explaining the way they were created and used, one could evoke or recreate an entire way of life (see 1942, 1957, 1967, 1992).

Evans's publications are still a pleasure to read. His *Irish Folk Ways* (1957), for example, provides an elegant account of landscapes, houses, farmyards, fences, carts, boats, weddings and seasonal customs, each set in a contemporary and historical context. He can be criticised when his emphasis on *vernacular* and neglect of *élite* culture becomes a distortion. Also, he skips too easily between historical periods, insufficiently noting that different historical periods are indeed different. Furthermore, like many of the Dublin-based folklorists, he often portrays Ireland as having an idyllic communal coherence with few social tensions. Nevertheless, his writings remain a *tour de force*, well worth revisiting. Evans did not however content himself with describing and drawing material objects. Instead, he turned to evoking the world of ordinary Ulster 'folk' by putting these same objects in a museum.

The project of establishing a folk museum had been mooted ever since the time of Bigger. Indeed, there had been a primitive folk museum in the Bigger-inspired toy-workshop in Ballycastle (McBrinn 2002, 45). Two world wars had disrupted the project. In 1953, however, Evans helped form the Committee on Ulster Folklife and Traditions, later to become the Ulster Folklife Society. Ronald Buchanan became editor of their new journal, *Ulster Folklife*, and in 1958, this Folklife Committee brought the Ulster Folk Museum[1] into existence (by Act of Parliament) under the directorship of George Thompson.

Thompson was (still is) a remarkable man. He had a feel for the lives of ordinary people and was a serious student of folklife, having done postgraduate work under Evans himself (Thompson 1958a, 1958b, 1982, Gailey *et al* 1964). He is remembered primarily for his qualities of leadership for it was largely due to his vision and quiet charisma that the Folk Museum flourished.

Alan Gailey, Keeper of Buildings and Thompson's successor as Director, was most immediately responsible for the discovery, research, dismantling and reassembling in the Folk Museum of vernacular buildings. Gailey also produced an outstanding body of writings. With one major exception and several small ones (1968, 1973b), this scholarship has been concerned with material culture. For example, he wrote about the spade (1970, 1982a), rope-twisters (1962b), furniture (1966), bonfires (1977), and harvest (1972a, 1973a, 1984a). Most extensively, he wrote about buildings (1961, 1962a, 1963, 1964, 1972b, 1974a, 1974b, 1984b, Gailey *et al* 1964). The major exception is a long association with the mummer's play (Gailey 1967, 1974c, 1975, 1979, 2005). His *Irish folk drama* (1969) remains the cornerstone of research into Irish mumming.

The strength of Gailey's scholarship, as that of Evans himself, lies in its use of the careful recording and locating of data to build a rounded picture of traditional Ulster. Gailey retained Evans's interest in survivals and diffusion, but he often expressed an affinity with the *Annales* school of French historiography, which he felt to be

1 Later the Ulster Folk and Transport Museum; subsequently subsumed into Museums and Galleries of Northern Ireland.

a bridge between his own form of ethnology and other emerging schools (Gailey 1990, 8). The sophistication of his methodology is found, for example, in his most recent article (2005) which uses a literary analysis to explore the genuine conundrum of the origin and development of the mumming play.

Evans had provided ethnologists in Ulster with a methodology and an example which allowed them to conjure an entire way of life out of the detailed study particularly of material objects. It was not surprising that those associated with the Ulster Folk Museum should continue to follow him even when they had ceased to accept his theoretical premises (see, for example, Carragher 1985, 1995, 1997, McManus 1984). Two of Evans's pupils, Ronald Buchanan and Desmond McCourt, maintained an especially close contact with the Museum. Buchanan published widely on different aspect of folklife (1955, 1956, 1957, 1962, 1963), and McCourt – who succeeded Buchanan as editor of *Ulster Folklife* – provided a stream of articles particularly on vernacular housing and field patterns (1955, 1956, 1962, 1965, 1970, Evans and McCourt 1968, 1971). A similar approach to building was later adopted by Philip Robinson (1976, 1977, 1979, 1982, 1985, 1986b, 1991) who assisted and then replaced Gailey as Keeper of Buildings and whose work was influenced by his earlier research into the Ulster Plantation (1984).

Paradigm Change: Rosemary Harris

Before 1970, a major problem with Ulster ethnology was that with its emphasis on objects and crafts, it had somehow sidestepped a central feature of Northern Irish life, its sectarianism. There had been exceptions. An early article by Mogey (1948) discusses the sectarian divide but without thinking it at all problematical or odd. Buchanan, who never accepted Evans's survivalism (1955), suggested that Ulster's sectarianism was down to the failure of Protestants and Catholics to intermarry (1956). This idea is an insight still battling to be understood. Again, the Ulster Folk Museum too – far ahead of its time – was quietly aware it might help overcome social division. In 1965, Evans himself wrote of the Museum, 'Here at least, in the effort to record, preserve, and study traditional Ulster ways and values a divided community appears to find common ground' (1965, 355). His sentiments were to be repeated by Thompson and his colleagues informally throughout the following decades (see Cashman 2001, 2006, Evans 1984). Despite these examples, until the 1970s, the quite substantial body of ethnological literature rarely mentioned, let alone studied Ulster's oh-so-important ethnic division. Then, suddenly, this elephant-in-the-living room became hyperactive, requiring that ideas already current in British, European and American ethnology be applied to Ulster.

Bronislaw Malinowski, founder of British social anthropology, had disputed the twin Darwinian notions of diffusionism and survivalism as long ago as the 1920s. It was not possible, he claimed, in the faraway societies where anthropologists worked, to reconstruct the past from objects and customs found in the present. Better to see these items as contributing to the functioning of society and therefore the satisfaction of human needs in the here and now.

By the 1950s, Malinowski's 'structural functionalism' was itself under fire. A number of overlapping and competing ideas – semiotics, structuralism, phenomenology, existentialism, Marxism and others – dramatically led ethnology away from both diffusionism/survivalism and structural-functionalism. Collingwood invited historians to put themselves into the minds of historical actors: to relive what Caesar *thought* as he crossed the Rubicon (1946, 213–217). Merton (1949) so clarified the concepts of functional sociology that people came to question whether 'society' could ever be a functioning whole. Lévi-Strauss (1958, 4ff) doubted that objects or customs belonged to 'types' or 'species'. Wittgenstein (1958) came to believe that words had meanings only when they were used in particular 'language games'. Berger and Luckmann (1967) showed that people lived according to 'socially constructed' beliefs. Kuhn (1962) thought natural science itself operated only through socially constructed *paradigms*. There was a shift, therefore, towards the idea that 'social reality' was composed of the subjective meanings that arose in social situations.

The book that transformed Ulster's ethnology was *Prejudice and tolerance in Ulster* by Rosemary Harris (1972), a former associate of Buchanan and pupil of Evans. This book did not show merely what the newspapers (and frequent explosions rattling one's windows) were now making plain, that Catholics and Protestants in Northern Ireland often had differing political ambitions. Nor did Harris just try to show they had different 'cultures'. Rather she showed that relations between Catholics and Protestants in Northern Ireland were subtle and complicated.

The population of the area around 'Ballybeg', she claimed, participated in two opposed ways of life. One was perceived as 'Catholic', but was in reality associated with the poorer farms in the hills. The other was perceived as 'Protestant' but was more properly associated with the more prosperous lowlands. Despite these perceptions, Catholics and Protestants lived both in the hills and in the valley. Catholics in the valley lived like 'Protestants', and Protestants in the hills lived like 'Catholics'. In each area, sectarian divisions and prejudices coexisted with a sense of shared community.

Harris had transformed Ulster's ethnology by relating Ulster's sectarian divide to social geography. She differed, however, from Evans in one important respect. She entirely explained the present in terms of the present. There was virtually no reference to the past.

Harris's study led immediately to similar work. Leyton described a different but similar pattern in the Mourne Mountains (1975a, 1975b). Here too, sectarian tension coexisted with a shared sense of community. McFarlane's remarkable but never-published study of gossip (1978) further demonstrated the importance of observing face-to-face social interactions. In Buckley's 'Upper Tullagh' (1982), Catholics and Protestants took care to remain on friendly terms. Here, however, stereotypes of Protestant and Catholic did not come from geographical residence. Rather the population drew on stereotypes of other relationships such as female/male; parent/child; employer/employee; rich/poor. Later, a sophisticated historical study by Akenson (1991) cemented the idea that sectarian stereotypes depended radically on local conditions.

These writers seemed partially to have rescued Ulster's communitarian ideals out of the wreckage of the Troubles. Donnan and McFarlane (1986), even worried they had gone too far. In practice, however, ethnology in Ulster was becoming deeply sombre. The rural idyll so often portrayed in earlier ethnology was crumbling.

Old Wine, New Bottles

Soon there were community studies based in urban settings. Ethnologists also returned to older topics, to seasonal customs, cures, folktales and the mummer's play. They used, however, newer, relativistic methodologies and concentrated on interpersonal interaction, on subjective meanings and, above all, on the Troubles.

Urban studies revealed a Northern Ireland where sectarian aggression mingled with poverty (for example, Burton 1978, Jenkins 1982, 1983, Howe 1990).[2] Few ethnologists after 1972 avoided making explicit or implicit political commentaries on the Troubles. Several studies got close to the brutality (Feldman 1991, Sluka 1989, 1999) or to the pathos (Santino 2001) of Northern Ireland life. Darby (1995), looking back to Simmel (1915), claimed that conflict – or at least ambivalence[3] – was at the heart of all human relationships.

It is typical of the new mood that studies of the well-established folklife topic of seasonal customs should now embrace sectarianism. The careful description that characterises Buchanan's still unsurpassed study of seasonal customs in Ulster (1962, 1963) gave way to studies by Robinson (1994) and Santino (1998a, 1998b) which explored the meanings these festivals might have in a time of conflict.

So too with the Twelfth of July. Gailey was the first professional ethnologist to look at events surrounding the Twelfth. His exemplary study of bonfires (1977) explains that most bonfires were (some still are) held on the night before certain quarter days, before the feast of St Peter and St Paul, and before the First and the Twelfth of July, as well as for other less regular purposes. Later studies that deal with the Twelfth shift their emphasis away from a careful interest in location, or indeed in the breadth of the tradition. Instead, they focus on the meaning of these events in relation to social conflict.

Jarman's studies of parades and visual displays (1997, 1998), for example, are heavy with the politics of defining urban spaces. Jarman and Bryan have written extensively on parades seeing them as 'a ritual expression of the ethnic differences which exist in Northern Ireland' (Bryan and Jarman 1997, 212). Several of their studies have included an orientation towards identifying appropriate public policies in relation to parades and symbolic displays (Bryan 1996, 2000, Bryan and Jarman 1996, 1999, Bryan *et al* 1995, Bryan and Gillespie 2005, Jarman 1978, 2002a, see also Donnan and McFarlane, 1989). Indeed, one can fairly say that this work has

2 The essays contained in *Irish Urban Cultures* (Curtin et al 1993) are unusual in that they discuss mostly the non-violent aspects of Ulster towns using the methodologies of urban anthropology.

3 The word is not Simmel's but actually Freud's (1915) and it is remarkable that these two men should publish such similar ideas in the same year.

made a major contribution to the resolution of these specific conflicts and to the Northern Ireland peace process generally.

There have been comparable studies with different perspectives. Buckley (1985–86) seeks to unpick the meaning of the Bible stories depicted on the banners of the Royal Black Institution. Bryan (1998) looks at the Twelfth through the eyes of journalists who find difficulty representing the dissonant coexistence of family festival and social disorder all in the same article, or even on the same page. Buckley and Kenney's description of riots (1995, 153–172) discovers a similar ambiguity, where family festival and violence, hymn singing and bottle throwing, fighting and fun all incongruously collide. Their book, *Negotiating Identity* (1995, see also Buckley 1998) more generally examines the processes involved in constructing identity in a variety of situations.

There have been several studies of banner paintings, some of which have emphasised the systematic similarities and differences between nationalist and Orange artistic traditions. Such studies often compare sectarian art with that found among such organisations such as the Freemasons and the trade unions (Loftus 1978, 1990, 1994, Jarman 1999). Buckley has taken this idea further, looking at the more general tradition of forming brotherhoods (Buckley, 1987, Anderson and Buckley 1988, Buckley 2000, 2007, see also Robinson 1986, Kilpatrick 1996).

Leaving aside the Troubles, one feature of the new ethnology has been the blurring of distinctions between disciplines. Bell's agricultural studies exemplify this change. His doctoral dissertation (1982) explicitly abandons the rigid distinction between history and anthropology (see also Bryan and Tonkin 1996). Bell's publications draw on his training as an anthropologist, while he writes, nevertheless, as a historian (1979, 1983, 1985, 1992, 2005, Bell and Watson 1986). Bell weaves into his historical narratives notions of 'tradition' that were central to Evans's ethnology. Above all, his studies of material culture are at one with his understanding of the social setting where objects are created and used.

Another agricultural specialist, Watson (1979, 1980, 1982, 1988, 2000, 2003) became particularly interested in agricultural *techniques*, in spades and swing ploughs, flachters, pigs and ponies. Here too, an ethnological approach merges with a concern for historical processes.

Contextualization also became important to folk-musicology, and this increasingly meant a concern with politics. Shields's considerable and valuable collection of traditional ballads (H. Shields 1964, 1971, 1974, 1981a, 1981b 1987, H. and L. Shields 1975) is firmly in the spirit of Edward Bunting, and it is entirely proper that traditional music of different kinds should be recorded and published. Feldman and Doherty (1979), however, not only give extensive transcriptions of fiddle music found in Tyrone and Donegal, but also offer a full account of the circumstances in which the music was played. Prosser (née Scullion) approaches fiddle music in the manner of a genuine ethnomusicologist (Scullion 1980) trying to discern the musicians' perceptions of their music. Her account of the Lambeg drum (Scullion 1981), and even more her description of a soloist singing 'The Craigywarren Heroes' to accompany an Orange ritual (Prosser 1982–85) introduces the reader to a very special, unfamiliar and intimate world.

Religion proper, long a topic for Irish historians, has received remarkably little attention from ethnologists. The studies that do exist concentrate on anti-Catholic forms of Protestantism. The most prolific writer, Bruce, is centrally a sociologist of religion (with research interests beyond Ulster), who has explored Paisleyism and political loyalism (Bruce 1986, 1992a, 1992b, 2007, Bruce *et al* 1986). Brewer and Higgins (1998, see also Buckley 1989) have explored the relation between Protestant theology and conflict. Buckley and Kenney (1995, 111–138) examined Pentecostalism, with an emphasis on conversion and religious experience. Buckley (1980) has also looked at the mostly religious practices he calls 'unofficial healing'.

Similar developments have taken place in the study of narrative. Glassie, who is not above making mere collections (1982b), more seriously grounds his discussion of folktales and historical narrative (1982a, 2006) in the relationship between story, storyteller and community. Jenkins (1977) sternly links fairy-belief to rural aggression and deviance. Ballard's (née Smith) work in folk-narrative is more literary, concerned with the nature of narrative. For example, she notes the strange identification of fairy belief with the Danes (Smith 1979), the way notions of authenticity often frame the Irish folktale (Ballard 1980) and, in an interesting discussion of seal stories, the quasi-totemic relationship between animal and community in Rathlin (Ballard 1983).[4]

There have been developments in the study of the mummer's play. I mentioned that Gailey provided the ground on which all subsequent studies of the Irish mummer's play have rested (1967, 1969, 1974c 1975, 1979, 2005). His work primarily examines texts and locates them in time and space with a view to tracing their history. In contrast, Glassie's studies of mumming (1976, 2007) are more humanistic (even sentimental) paying attention less to the texts themselves than to their social context. Revisiting the topic he (2007) and also Cashman (2000, 2007) show how this drama has had a subtle and complex role in healing torn relationships between Catholic and Protestant.

Occasionally (in the circumstances, remarkably rarely), ethnology has become controversial. This happened with language. There had been research into Ulster dialect forms of English from the 1960s, largely through the work of Brendan Adams, Braidwood and Gregg (Barry *et al* 1982, Adams *et al* 1964, Braidwood 1965, 1966, 1971, 1972, 1975, 1978, Montgomery *et al* 2006 139–267). There arose, however, a claim that the 'Ulster Scots' speech found in north-west and north-east Ulster was a separate 'minority language'. Some commentators construed this claim as a politically motivated attempt to counter the Irish language espoused by nationalists, as a move in what Harrison has called 'symbolic conflict'.[5] One can gain a flavour of this debate by looking at the two editions of *Ulster Folklife* (44 and 45) largely devoted to it.

---

4 Ballard has also worked extensively on textiles (1988, 1989, 1992, 1994) and marriage (1998).

5 In one of the most interesting publications to arise indirectly out of the Troubles, Harrison (1995) shows how social groups fight by creating new culture; by overwhelming or stealing the culture of their opponents, or by claiming that their own culture has the higher value.

Conclusion

Studies of such a small population have been remarkably many and rich, fuelled, as was inevitable, by the ferocious conflict that broke out in the late 1960s. That conflict precipitated a major paradigm-change among Ulster's ethnologists. The best of these later writers now rise above Collingwood's 'externals' to discover 'the ideas, perceptions, values, beliefs and assumptions which are the basis of peoples' lifestyles and practices' (Donnan and McFarlane 1989, 3).

I have taken care, however, not to suggest that earlier generations of Ulster ethnologists were mistaken while the later ones were correct. Since the 1970s, there has grown a heightened awareness that people continually transform ideas, objects and practices so they have different meanings in different contexts. Nevertheless, empirical inquiry still remains central to ethnology. It is still useful to suppose that ideas and practices survive and get diffused from place to place, and that certain features of social life sustain a social group while others do not. Even in this article, I have spoken of ethnological 'traditions': I am aware that many of my own views have 'survived' from the days of Malinowski, or were 'diffused' from France and America. And if ethnologists no longer portray rural idylls, they have shown that community is still a feature of human life even in the worst of circumstances. There is, therefore, a dialectical process, according to which each generation rejects old ideas only to rediscover these same ideas in due course. The work of the Dublin folklorists and of Evans and his successors remain impressive. Even Newton knew that he stood on the shoulders of giants. We should remember that we do too.

Bibliography

Abraham, A.S.K. (2000) 'The T.G.F. Paterson manuscript collection at Armagh County Museum,' *Ulster Folklife*, 46, 42–47.

Adams, B., Braidwood, J. and Gregg, R.J. (eds) (1964) *Ulster Dialects: An Introductory Symposium*. Holywood: Ulster Folk and Transport Museum.

Adams, R. (1985) 'Ulster folklife 1738–1740 from the pages of the Belfast Newsletter', *Ulster Folklife*, 31, 41–52.

Akenson, D. (1991) *Small Differences: Irish Catholics and Irish Protestants, 1815–1922*. Dublin: Gill and Macmillan.

Anderson, K. and Buckley, A.D. (1988) *Brotherhoods in Ireland*. Cultra: Ulster Folk and Transport Museum.

Andrews, J. (1975) *A Paper Landscape: The Ordnance Survey of Ireland*. Oxford: Clarendon.

Audley, B. (2003) 'Some Missing Items of the Bunting Collection rediscovered', *Ulster Folklife*, 49, 1–5.

Ballard, L. (1994) 'Aspects of the History and Development of Irish Dance Costume', *Ulster Folklife*, 40, 62–67.

—— (1980) 'Ulster Oral Narrative: The Stress on Authenticity', *Ulster Folklife*, 26, 35–40.

—— (1983) 'Seal Stories and Belief on Rathlin Island', *Ulster Folklife*, 29, 33–42.

—— (1988) 'Some Photographic Evidence of the Practice of Dressing Boys in Skirts', *Ulster Folklife*, 34, 41–7.
—— (1989) 'The Lilley Collection', *Ulster Folklife*, 35, 8–18.
—— (1992) 'Some Aspects of Tradition among Female Religious in Ulster', *Ulster Folklife*, 38, 68–78.
—— (1998) *Forgetting Frolic: Marriage Traditions in Ireland.* Belfast: Institute of Irish Studies Queen's University of Belfast.
Barry, M. and Tilling, P. (eds) (1982) *The English Dialects of Ulster: An Anthology of Articles on Ulster Speech by G.B. Adams.* Cultra: Ulster Folk and Transport Museum.
Bell, J. (1982) *Economic Change in the Dunfanaghy Area of North Donegal, 1990–1940.* Unpublished Doctoral dissertation. Belfast: Queen's University of Belfast.
—— (1988) 'Intelligent revivalism: the First *Feis na nGleann*, 1904', in A. Gailey (ed.), *The Use of Tradition: Essays Presented to G.B. Thompson.* Cultra: Ulster Folk and Transport Museum, pp. 3–12.
—— (1979) 'Hiring Fairs in Ulster', *Ulster Folklife*, 25, 67–78.
—— (1983) 'The Use of Oxen on Irish Farms since the Eighteenth Century', *Ulster Folklife*, 29, 18–28.
—— (1985) 'Farm Servants in Ulster', *Ulster Folklife*, 31, 13–20.
—— (1992) *People and the Land: Farming Life in Nineteenth Century Ireland.* Belfast: Friar's Bush.
—— (1998) 'Concepts of survival and revival in Irish culture', *Ulster Folklife*, 44, 100–109.
—— (2005) *Ulster Farming Families, 1930–1980.* Belfast: Ulster Historical Foundation.
Bell, J. and Watson, M. (1986) *Irish Farming: Implements and Techniques 1750–1900*. Edinburgh: John Donald.
Berger, P. and Luckmann, T. (1967) *The Social Construction of Reality: A Treatise in the Sociology of Knowledge*. London: Allen Lane.
Braidwood, J. (1965) 'Local Bird Names in Ulster – A Glossary', *Ulster Folklife*, 11, 98–135.
—— (1966) 'Local Bird Names in Ulster – A Glossary', *Ulster Folklife*, 12, 104–107.
—— (1971) Local Bird Names in Ulster – A Glossary', *Ulster Folklife*, 17, 81–84.
—— (1972) 'Terms for "Left-handed" in the Ulster Dialects', *Ulster Folklife*, 18, 98–110.
—— (1975) *The Ulster Dialect Lexicon.* Inaugural lectures. Belfast: Queen's University of Belfast.
—— (1978) 'Local Bird Names in Ulster: Further Additions', *Ulster Folklife*, 24, 83–87.
Brewer, J. with Higgins, G. (1998) *Anti-Catholicism in Northern Ireland 1600–1998.* Basingstoke: Macmillan.
Briody, M. (2005) '"Publish or perish": the vicissitudes of the Irish Folklore Institute', *Ulster Folklife*, 51, 10–33.
Bruce, S. (1986) *God Save Ulster: The Religion and Politics of Paisleyism*. Oxford: Clarendon Press.

Bruce, S. (1992a) *Northern Ireland: Reappraising Loyalist Violence.* Belfast: Research Institute for the Study of Conflict and Terrorism.

—— (1992b) *The Red Hand: Protestant Paramilitaries in Northern Ireland.* Oxford: Oxford University Press.

—— (2007) *Paisley: Religion and Politics in Northern Ireland.* Oxford: Oxford University Press.

Bruce, S., Taylor, R. and Wallis, R. (1986) *No Surrender: Paisleyism and the Politics of Ethnic Identity in Northern Ireland.* Belfast: Department of Social Studies, Queen's University of Belfast.

Bryan, D. (1996) *Ritual, 'Tradition' and Control: The Politics of Orange Parades in Northern Ireland.* Coleraine: University of Ulster.

—— (1998) '"Ireland's very own Jurassic Park" the mass media, Orange parades and the discourse on tradition', in A.D. Buckley (ed.), *Symbols in Northern Ireland.* Belfast: Institute of Irish Studies, Queen's University of Belfast, pp. 23–42.

—— (2000) *Orange Parades: The Politics of Ritual, Tradition and Control.* London: Pluto.

Bryan, D. and Gillespie, G. (2005) *Transforming Conflict: Flags and Emblems.* Belfast: Institute of Irish Studies, Queen's University of Belfast.

Bryan, D. and Jarman, N. (1996) *Parade and Protest: A Discussion of Parading Disputes in Northern Ireland.* Coleraine: Centre for the Study of Conflict, University of Ulster.

—— (1997) 'Parading tradition, protesting triumphalism: utilising anthropology in public policy' in H. Donnan and G. McFarlane (eds), *Culture and Policy in Northern Ireland.* Belfast: Institute of Irish Studies, Queen's University of Belfast, pp. 211–29.

—— (1999) *Independent Intervention: Monitoring the Police, Parades and Public Order.* Belfast: Democratic Dialogue and the Community Development Centre.

Bryan, D. and Tonkin, E. (1996) 'Political ritual: temporality and tradition' in A. Boholm (ed.), *Political Ritual.* Gothenberg: IASSA, pp. 14–36.

Bryan, D., Fraser, T.G. and Dunn, S. (1995) *Political Rituals: Loyalist Parades in Portadown.* Coleraine: University of Ulster, Centre for the Study of Conflict.

Buchanan, R. (1955) 'The study of folklore', *Ulster Folklife*, 1, 8–12.

—— (1956) 'The folklore of an Irish townland', *Ulster Folklife*, 2, 43–55.

—— (1957) 'Stapple Thatch', *Ulster Folklife*, 3, 19–28.

—— (1962) 'Calendar customs, I', *Ulster Folklife*, 8, 15–34.

—— (1963) 'Calendar customs, II', *Ulster Folklife*, 9, 61–79.

—— (2004) 'Editorial', *Ulster Folklife*, 50, 1–3.

Buckley, A.D. (1980) 'Unofficial healing in Ulster', *Ulster Folklife*, 26, 15–34.

—— (1982) *A Gentle People: A Study of a Peaceful Community in Northern Ireland.* Cultra: Ulster Folk and Transport Museum.

—— (1985–86) '"The chosen few": Biblical texts in the regalia of an Ulster secret society', *Folklife*, 29, 15–24.

—— (1987) '"On the club": Friendly Societies in Ireland', *Irish Economic and Social History*, 14, 39–58.

—— (1989) '"We're trying to find our identity": uses of history among Ulster Protestants' in E. Tonkin, M. McDonald and M. Chapman (eds), *History and Ethnicity*. London: Routledge, pp. 183–91.

—— (1998) 'Introduction: Daring us to laugh: creativity and power in Northern Irish symbols' in A.D. Buckley (ed.), *Symbols in Northern Ireland*. Belfast: Institute of Irish Studies, Queen's University of Belfast, pp. 1–21.

—— (2000) 'Royal Arch, Royal Arch Purple and Raiders of the Lost Ark: Secrecy in Orange and Masonic ritual' in T. Owen (ed.), *From Corrib to Cultra: Folklife Essays in Honour of Alan Gailey*. Belfast: Institute of Irish Studies, Queen's University Belfast with Ulster Folk and Transport Museum, pp. 163–80.

—— (2007) '"Rise up dead man, and fight again": Mumming, the Mass and the Masonic Third Degree' in A.D. Buckley, S. Ó Catháin, C. Mac Cárthaigh and S. Mac Mathúna (eds), *Border-Crossing: Mumming in Cross-Border and Cross-Community Contexts*. Dundalgan Press: Dundalk, pp. 19–38.

Buckley, A.D. and Kenney, M.C. (1995) *Negotiating Identity: Rhetoric, Metaphor and Social Drama in Northern Ireland*. Washington: Smithsonian Institution Press.

Buckley, A.D., Ó Catháin, S., Mac Cárthaigh, C. and Mac Mathúna, S. (eds) (2007) *Border-Crossing: Mumming in Cross-Border and Cross-Community Contexts*. Dundalk: Dundalgan Press.

Burton, F. (1978) *The Politics of Legitimacy: Struggles in a Belfast Community*. London: Routledge.

Carlton, W. (1832) *Traits and Stories of the Irish Peasantry*. Dublin: Curry.

Carlyle, T. (1849) *Reminiscences of my Irish Journey in 1849*. London: S Low, Marston, Searle and Rivington (1882).

Carragher, F. (1985) 'Settle Beds in the Ulster Folk and Transport Museum', Ulster *Folklife*, 31, 36–40.

Carragher, F. (1995) 'Cluan Place, Ballymacarrett, Belfast', *Ulster Folklife*, 41, 1–11.

—— (1997) 'Miss Margaret Clyde of Duncrun', *Ulster Folklife*, 43, 48–57.

Cashman, R. (2000) 'Mumming with neighbours in west Tyrone', *Journal of Folklore Research*, 37, 73–84.

——(2001) 'Can history heal? The uses of local history in a Northern Irish border community', *Irish Studies Working Papers*, 1, Florida: Nova Southeastern University, 5–11.

—— (2006) 'Critical nostalgia and material culture in Northern Ireland', *Journal of American Folklore*, 119, 137–60.

—— (2007) 'Mumming on the Irish border: social and political implications', in A.D. Buckley, S. Ó Catháin, C. Mac Cárthaigh and S. Mac Mathúna (eds) (2007) *Border-crossing: Mumming in Cross-border and Cross-community Contexts*. Dundalk: Dundalgan Press, pp. 39–56.

Collingwood, R.G. (1946) *The Idea of History*, Oxford, Clarendon (1961).

Curtin, C., Donnan, H. and Wilson, T.M. (eds) (1993) *Irish Urban Cultures*. Belfast: Institute of Irish Studies Queen's University of Belfast.

Darby, J. (1995) *What's Wrong with Conflict?* Coleraine: University of Ulster.

Dixon, R. (1997) 'Francis Joseph Bigger: Belfast's cultural Don Quixote', *Ulster Folklife*, 43, 40–57.

Donnan, H. and McFarlane, G. (1986) 'Social anthropology and the sectarian divide in Northern Ireland' in R. Jenkins, H. Donnan and G. McFarlane (eds), *The Sectarian Divide in Northern Ireland Today*. Occasional papers 41, London: Royal Anthropological Institute of Great Britain and Ireland, pp. 23–37.

—— (1989) 'Introduction' in H. Donnan, and G. McFarlane (eds), *Social Anthropology and Public Policy in Northern Ireland*. Aldershot: Avebury pp. 1–25.

Evans, E. and Gafkin, M. (1935) *Belfast Naturalists' Field Club Survey of Antiquities: Megaliths and Raths*. Belfast.

Evans, E. (1942) *Irish Heritage: The Landscape, the People and their Work*. Dundalk: Tempest.

—— (1965) 'Folklife studies in Northern Ireland', *Journal of the Folklore Institute*, 2–3, 355–65.

—— (1966) *Prehistoric and Early Christian Ireland: A Guide*. London: Batsford.

—— (1967) *Mourne Country: Landscape and Life in South Down*. Dundalk: Dundalgan.

—— (1971) 'Thomas George Farquhar Paterson OBE, MA, MIRA', *Ulster Journal of Archaeology* 34, 1–2.

—— (1957) *Irish Folk Ways*. London: Routledge and Kegan Paul.

—— (ed.) (1975) *Harvest Home: The Last Sheaf: A Selection from the Writings of TGF Paterson*. Armagh: Armagh County Museum.

—— (1984) *Ulster: The Common Ground*. Mullingar: Lilliput.

—— (1988) The early development of folklife studies in Northern Ireland in A. Gailey (ed.), *The Use of Tradition: Essays Presented to G.B. Thompson*. Cultra: Ulster Folk and Transport Museum, pp. 91–96.

—— (1992) *The Personality of Ireland: Habitat, Heritage and History*. Dublin: Lilliput.

Evans, E. and McCourt, D. (1968) 'A Late Seventeenth-Century Farmhouse at Shantallow, near Londonderry. Part 1', *Ulster Folklife*, 14, 4–23.

—— (1971) 'A Late Seventeenth-Century Farmhouse at Shantallow, near Londonderry. Part 2', *Ulster Folklife*, 17, 37–41.

Feldman, A. and O'Doherty, E. (1979) *The Northern Fiddler: Music and Musicians of Donegal and Tyrone*. Belfast: Blackstaff.

Feldman, A. (1991) *Formations of Violence: The Narrative of the Body and Political Terror in Northern Ireland*. Chicago: University of Chicago Press.

Fleure, H.J. (1947) *Some Problems of Society and Environment: Three Lectures Delivered at University College London*. Liverpool: Philip.

Freud, S. (1915) 'Instincts and their vicissitudes' in *The Standard Edition of the Complete Psychological Works of Sigmund Freud, Volume XIV (1914–1916)*, Translated under the editorship of James Stachey in collaboration with Anna Freud. London: Vintage (2001), pp. 109–40.

Gailey, A. (1961) 'The Thatched houses of Ulster', *Ulster Folklife*, 7, 9–18.

—— (1962a) 'Two cruck truss houses near Lurgan', *Ulster Folklife*, 8, 57–64.

—— (1962b) 'Ropes and rope-twisters', *Ulster Folklife*, 8, 72–82.

—— (1963) 'The cots of North Derry', *Ulster Folklife*, 9, 46–52.
—— (1964) 'Notes on three cruck-truss houses', *Ulster Folklife*, 10, 88–94.
—— (1966) 'Kitchen furniture', *Ulster Folklife*, 12, 18–31.
—— (1967) 'The rhymers of South-east Antrim', *Ulster Folklife*, 13, 18–28.
—— (1968) 'Edward L Sloan's "The Year's Holidays"', *Ulster Folklife*, 14, 51–59.
—— (1969) *Irish folk drama*. Cork: Mercier Press.
—— (1970) 'The typology of the Irish spade' in A. Gailey (ed.), *The Spade in Northern and Atlantic Europe.* Belfast: Ulster Folk Museum and Institute of Irish Studies Queen's University of Belfast, pp. 35–39.
—— (1972a) 'The last sheaf in the north of Ireland', *Ulster Folklife*, 18, 1–79.
—— (1972b) 'Further cruck-trusses in East Ulster', *Ulster Folklife*, 18, 80–97.
—— (1973a) 'The flax harvest', *Ulster Folklife*, 19, 24–29.
—— (1973b) 'Sources for the Historical Study of Easter as a Popular Holiday in Ulster', *Ulster Folklife*, 26, 68–74.
—— (1974a) 'A house from Gloverstown, Lismacloskey, Co Antrim', *Ulster Folklife*, 20, 24–41.
—— (1974b) 'The housing of the rural poor in nineteenth century Ulster', *Ulster Folklife*, 22, 34–58.
—— (1974c) 'Chapbook influence on Irish mummers' plays', *Folklore*, 85, 1–22.
—— (1975) 'The Christmas rhyme', *Ulster Folklife*, 21, 73–84.
—— (1977) 'The bonfire in North Irish tradition', *Folklore*, 88, 3–38.
—— (1979) 'Mummers' and Christmas rhymers' plays in Ireland: the problem of distribution', *Ulster Folklife*, 24, 59–68.
—— (1982a) *Spade Making in Ireland*, Holywood. Ulster Folk and Transport Museum.
—— (1982b) 'Folk-life study and the Ordnance Survey Memoirs' in A. Gailey and D Ó hOgáin (eds), *Gold Under the Furze: Studies in Folk Tradition Presented to Caoimhin Ó Danachair.* Dublin: Glendale, pp. 154–64.
—— (1982c) 'Folk culture, context and social change' in E. Hörander, and H. Lunzen *Folklorismus.* Neusiedl, pp. 73–102.
—— (1984a) 'Introduction and spread of the horse-powered threshing machine to Ulster's farms in the nineteenth century', *Ulster Folklife*, 30, 37–54.
—— (1984b) *Rural houses in the north of Ireland.* Edinburgh: Donald.
—— (1990) '... such as pass by us daily ... the study of folklife', The Estyn Evans Lecture (1989), in *Ulster Folklife*, 36, 4–22.
—— (2005) 'Chapbook printings of mummer's plays in Ireland', *Ulster Folklife*, 51–53.
Gailey, A., McCourt, D. and Thompson, G. (1964) 'The Magilligan cottier house', *Ulster Folklife*, 10, 23–34.
Glassie, H. (1976) *All Silver and No Brass: An Irish Christmas Mumming*. Dublin: Dolmen.
—— (1982a) *Passing the Time in Ballymenone: Culture and History of an Ulster Community.* Pennsylvania: University of Pennsylvania Press.
—— (1982b) *Irish Folk History: Folktales from the North.* Dublin: O'Brien Press.
—— (2006) *The Stars of Ballymenone.* Bloomington: Indiana University Press.

—— (2007) 'Mumming in Ballymenone', in A.D. Buckley, S. Ó Catháin, C. Mac Cárthaigh and S. Mac Mathúna (eds), *Border-crossing: Mumming in Cross-border and Cross-community Contexts*. Dundalk: Dundalgan Press, pp. 91–101.

Hall, Mr and Mrs S.C. (1850) *Ireland, its Scenery, Character etc.* London, Virtue (3 vols).

Harbison, J. (1989) 'The Belfast Harpers' Meeting, 1792: the legacy,' *Ulster Folklife*, 35, 113–28.

Harris, R. (1972) *Prejudice and Tolerance in Ulster: A Study of Neighbours and Strangers in a Border Community.* Manchester, Manchester University Press.

Harrison, S. (1995) 'Four types of symbolic conflict', *Journal of the Royal Anthropological Institute*, 1, 255–73.

Howe, L. (1990) *Being Unemployed in Northern Ireland: An Ethnographic Study*. Cambridge: Cambridge University Press.

Jarman, N. (1997) *Material Conflicts: Parades and Visual Displays in Northern Ireland.* Oxford: Berg.

—— (1998) 'Painting landscapes: the place of murals in the symbolic construction of urban space', in A.D. Buckley (ed.), *Symbols in Northern Ireland.* Belfast: Institute of Irish Studies, Queen's University of Belfast, pp. 81–98.

—— (1999) *Displaying Faith: Orange, Green and Trade Union Banners in Northern Ireland*. Belfast: Institute of Irish Studies, Queen's University of Belfast.

—— (2002a) *Managing Disorder: Responding to Interface Violence in North Belfast.* Belfast: Office of the First Minister and Deputy First Minister.

Jenkins, R. (1977) 'Witches and fairies: supernatural aggression and deviance among the Irish peasantry', *Ulster Folklife*, 23, 33–56.

—— (1982) *Hightown Rules: Growing up in a Belfast Housing Estate.* Leicester: National Youth Bureau.

—— (1983) *Lads, Citizens and Ordinary Kids: Working-class Youth Lifestyles in Belfast.* London: Routledge.

Kilpatrick, C. (1996) 'Black, Scarlet, Blue, Royal Arch Purple, Or Any Other Colour', *Ulster Folklife*, 42, 23–31.

Kuhn, T. (1962) *The Structure of Scientific Revolutions*. Chicago: University of Chicago Press.

Lévi-Strauss, C. (1958) 'Introduction: History and Anthropology', *Structural Anthropology, 1*. Translated by C Jacobson and B G Schoepf. London: Penguin (1993) pp. 1–27.

Leyton, E. (1975a) *The One Blood: Kinship and Class in an Irish Village.* Newfoundland: St. John's, Institute of Social and Economic Research, Memorial University of Newfoundland.

—— (1975b) 'Opposition and integration in Ulster', in *Man*, 9, 185–98.

Locke, J. (1859) 'On the heath beer of the ancient Scandinavians', *Ulster Journal of Archaeology*, 7, 219–26.

Loftus, B. (1978) Marching workers: an exhibition of Irish trade banners and regalia (held at the) Ulster Museum, Belfast, 2–28 May. Belfast and Dublin: Arts Council and Arts Council of Northern Ireland.

—— (1990) *Mirrors: William III & Mother Ireland.* Dundrum: Picture Press.

Loftus B. with Atkinson, C. (1994) *Mirrors: Orange and Green*. Dundrum, Co. Down: Picture Press.

Maloney, C. (2000) *The Irish Music Manuscripts of Edward Bunting (1773–1843): An Introduction and Catalogue*. Dublin: Irish Traditional Music Archive.

MacAdam, R. (1859) 'Six hundred Gaelic proverbs collected in Ulster (continued)', *Ulster Journal of Archaeology*, 7, 278–87.

McBrinn, J. (2002) 'The peasant and folk art revival in Ireland, 1890–1920 with special reference to Ulster', *Ulster Folklife*, 48, 14–61.

—— (2004) 'The Princess Taise Banner at the 1904 Feis na nGleann', *Ulster Folklife*, 50, 71–98.

McCourt, D. (1955) 'Infield and Outfield in Ireland' *Economic History Review*, 7, 369–76.

—— (1956) 'The Outshot House-Type and its distribution in County Londonderry', *Ulster Folklife*, 2, 27–55.

—— (1962) 'Weavers' Houses around South-West Lough Neagh', *Ulster Folklife*, 8, 43–56.

—— (1965) 'Some Cruck-framed Buildings in Donegal and Derry', *Ulster Folklife*, 11, 39–50.

—— (1970) 'The House with Bedroom over Byre: A Long-house Derivative?' *Ulster Folklife*, 15/16, 3–27.

Megan McManus (1984) 'Coarse Ware' in D.S. Smith (ed.), *Ireland's Traditional Crafts*. London: Thames and Hudson, pp. 186–90.

McFarlane, G. (1978) *Gossip and social relations in a Northern Irish village* Unpublished doctoral dissertation. Belfast: Queen's University of Belfast.

Merton, R.K. (1949) *Social Theory and Social Structure*. New York: Free Press, Collier Macmillan (1969).

Mogey, J. (1948) 'The community in Northern Ireland', *Man*, 48, 85–87.

Montgomery, M., Robinson, P. and Smyth, A. (2006) *The Academic Study of Ulster Scots: Essays for and by Robert J. Gregg*. Belfast: National Museum and Galleries of Northern Ireland.

Murphy, M.J. (1964) *Mountain Year*. Oxford: Oxford University Press.

—— (1965) 'The Folk Stories of Dan Rooney of Lurgancanty', *Ulster Folklife*, 11, 80–86.

—— (1973) 'Folktales and Traditions from County Cavan and South Armagh', *Ulster Folklife*, 19, 30–37.

—— (1975) *Now You're Talking – Folk Tales from the North of Ireland* Belfast: Blackstaff Press.

—— (1976) *Mountainy Crack: Tales of Slieve Gullioners*, Belfast: Blackstaff Press.

—— (1989) *My Man Jack: Bawdy Tales from Irish folklore*. Dingle: Brandon.

O'Laverty, J. (1859) 'Remarkable correspondence of Irish, Greek and Oriental legends', *Ulster Journal of Archaeology* 7, 334–46.

Ordnance Survey Memoirs (1835–40) *The Ordnance Survey Memoirs, 1835–1840*. Transcribed at the Institute of Irish Studies, Queen's University of Belfast under the direction of Angélique Day.

O'Sullivan, D. with Ó Súilleabháin, M. (1983) *Bunting's Ancient Music of Ireland.* Cork: Cork University Press.

Ó Tuathail, É. (1933) *Sgéalta Mhuintir Luinigh. Munterloney Folk-tales.* Dublin: Irish Folklore Institute.

Plumptree, A. (1817) *Narrative of a Resident of Ireland.* London: Colburn.

Prosser, F. (1982–5) '"The Heroes' Song" in an Orange ceremony', *Irish Folk Music Studies*, 4, 45–54.

Robinson, P. (1976) 'Irish Settlement in Tyrone before the Ulster Plantation', *Ulster Folklife*, 22, 59–69.

—— (1977) 'The Spread of Hedged Enclosure in Ulster', *Ulster Folklife*, 23, 57–69.

—— (1979) 'Vernacular Housing in Ulster in the Seventeenth Century', *Ulster Folklife*, 25, 1–28.

—— (1982) 'A Water-Mill built in 1615 by the Drapers' Company at Moneymore, County Londonderry', *Ulster Folklife*, 28, 49–55.

—— (1984) *The plantation of Ulster: British Settlement in an Irish Landscape 1660–1670.* Dublin: Gill and Macmillan.

—— (1985) 'From Thatch to Slate: Innovation in Roof Covering Materials of Traditional Houses in Ulster', *Ulster Folklife*, 31, 21–35.

—— (1986a) 'Hanging Ropes and Buried Secrets', *Ulster Folklife*, 32, 3–15.

—— (1986b) 'A Message for the Future: Note on a Building Custom', *Ulster Folklife*, 32, 48–53.

—— (1991) 'The Use of the Term "Clachan" in Ulster', *Ulster Folklife*, 37, 30–35.

—— (1994) 'Harvest, Halloween and Hogmanay: acculturation in some calendar customs of the Ulster Scots', in J. Santino (ed.), *Halloween and Other Festivals of Death and Life.* Knoxville: University of Tennessee Press, pp. 3–23.

Santino, J. (1998a) *The Hallowed Eve: Dimensions of Culture in a Calendar Festival in Northern Ireland.* Lexington: University Press of Kentucky.

—— (1998b) 'Light up the sky: Halloween bonfires and cultural hegemony in Northern Ireland' in A.D. Buckley (ed.), *Symbols in Northern Ireland*, Belfast: Institute of Irish Studies, Queen's University of Belfast, pp. 63–80.

—— (2001) *Signs of War and Peace: Social Conflict and the Use of Public Symbols in Northern Ireland.* New York: Palgrave.

Scullion, F. (1980) 'Perceptions of style amongst Ulster fiddlers' in *Studies in Traditional Music and Dance: Proceedings of the 1980 Conference.* London: UK National Committee (International Folk music council, 33–46.

—— (1981) 'History and Origins of the Lambeg Drum', *Ulster Folklife*, 27, 19–38.

Shields, H. (1964) 'Some Bonny Female Sailors', *Ulster Folklife*, 10, 35–45.

—— (1971) 'Some '"Songs and Ballads in use in the Province of Ulster 1845"', *Ulster Folklife*, 17, 3–24.

—— (1974) 'Some "Songs and Ballads in use in the Province of Ulster ... 1845"', *Ulster Folklife*, 18, 34–65.

—— (1981a) 'A Singer of Songs: Jimmy McCurry of Myroe', *Ulster Folklife*, 27, 1–18.

—— (1981b) *Rose and Thistle: Folk Singing in North Derry.* Belfast: Blackstaff.

—— (1987) 'Popular Broadsides in the Library of the Royal Society of Antiquaries of Ireland', *Ulster Folklife*, 33, 1–25, 24–65.

Shields, H. and L. (1975) 'Irish Folk-Song Recordings 1966–1972: an index of tapes in the Ulster Folk and Transport Museum', 21, 25–54.

Simmel, G. (1915) *Conflict: The Web of Group Affiliations*. New York: Free Press of Glencoe (1972).

Sluka, J. (1989) *Hearts and Minds, Water and Fish: Support for the IRA and INLA in a Northern Irish Ghetto*. London: AAI Press.

—— (1999) ''For God and Ulster': the culture of terror and loyalist death squads in Northern Ireland' in J. Sluka (ed.), *Death Squad: The Anthropology of State Terror*. Philadelphia: University of Pennsylvania Press, pp. 127–57.

Smith, L.-M. (1979) 'The Position of the "Danes" in contemporary Ulster oral narrative', *Ulster Folklife*, 25, 103–12.

Thompson, G. (1958a) *Primitive Land Transport of Ulster*. Belfast: Belfast Museum and Art Gallery.

—— (1958b) 'The blacksmith's craft', *Ulster Folklife*, 4, 33–36.

—— (1982) 'Applied folk-life study: a personal view', in A. Gailey and D. Ó hOgáin (eds), *Gold under the Furze: Studies in Folk Tradition Presented to Caoimhin Ó Danachair*. Dublin: Glendale, pp. 43–49.

Thackeray, W. (1843) 'The Irish sketch book' in W. Thackeray (1972) *The Paris Sketch Book, the Irish Sketch Book, and Notes of a Journey from Cornhill to Grand Cairo*. London: Smith, Elder and Co. pp. 259–558.

Tylor, E. (1871) *Primitive Culture: Researches into the Mythology, Philosophy, Religion, Art and Custom*. London: Murray.

Watson, M. (1979) 'Flachters: their construction and use in a Ulster peat bog,' in *Ulster Folklife*, 25, 61–66.

—— (1980) 'Cushendall hill ponies', *Ulster Folklife*, 26, 8–14.

—— (1982) 'North Antrim swing ploughs: their construction and use', *Ulster Folklife*, 28, 13–23.

—— (1988) 'Standardisation of farming practice: the case of the large white Ulster pig', *Ulster Folklife*, 34, 1–15.

—— (2000) 'Spades versus ploughs: a nineteenth-century debate Part 1', *Ulster Folklife*, 46, 48–68.

—— (2003) 'Spades and ploughs: a nineteenth-century debate. Part 2', *Ulster Folklife*, 49, 50–77.

Wittgenstein, L. (1958) *Philosophical Investigations*. Translated by G.E.M. Anscombe. New York: Macmillan.

Index